Nilpotence and Periodicity in Stable Homotopy Theory

by

Douglas C. Ravenel

PRINCETON UNIVERSITY PRESS

———

PRINCETON, NEW JERSEY

1992

Printed in the United States of America

The Annals of Mathematics Studies are edited by
Luis A. Caffarelli, John N. Mather, and Elias M. Stein

Princeton University Press books are printed on acid-free paper,
and meet the guidelines for permanence and durability of the
Committee on Production Guidelines for Book Longevity of the
Council on Library Resources

Library of Congress Catalog-in-Publication Data

Ravenel, Douglas C.
Nilpotence and periodicity in stable homotopy theory /
by Douglas C. Ravenel.
p. cm. — (Annals of mathematics studies ; no. 128)
Includes bibliographical references and index.
ISBN 0-691-08792-X (CL)—ISBN 0-691-02572-X (PB)
1. Homotopy theory. I. Title. II. Series.
QA612.7.R38 1992
514'.24—dc20   92-26785

Annals of Mathematics Studies

Number 128

To my children,
Christian, René,
Heidi and Anna

Contents

Preface

This research leading to this book began in Princeton in 1974–75, when Haynes Miller, Steve Wilson and I joined forces with the goal of understanding what the ideas of Jack Morava meant for the stable homotopy groups of spheres. Due to widely differing personal schedules, our efforts spanned nearly 24 hours of each day; we met during the brief afternoon intervals when all three of us were awake. Our collaboration led to [MRW77] and Morava eventually published his work in [Mor85] (and I gave a broader account of it in my first book, [Rav86]), but that was not the end of the story.

I suspected that there was some deeper structure in the stable homotopy category itself that was reflected in the pleasing algebraic algebraic patterns described in the two papers cited above. I first aired these suspicions in a lecture at the homotopy theory conference at Northwesern University in 1977, and later published them in [Rav84], which ended with a list of seven conjectures. Their formulation was greatly helped by the notions of localization and equivalence defined by Bousfield in [Bou79b] and [Bou79a].

I had some vague ideas about how to approach the conjectures, but in 1982 when Waldhausen asked me if I expected to see them settled before the end of the century, I could offer him no assurances. It was therefore very gratifying to see all but one of them proved by the end of 1986, due largely to the seminal work of Devinatz, Hopkins and Smith, [DHS88]. *The mathematics surrounding these conjectures and their proofs is the subject of this book.*

The one conjecture of [Rav84] *not* proved here is the telescope conjecture (7.5.5). I disproved a representative special case of it in 1990; an outline of the argument can be found in [Rav92]. I find this development equally satisfying. If the telescope conjecture had been proved, the subject might have died. Its failure leads to interesting questions for future work. On the other hand, had I not believed it in 1977, I would not have had the heart to go through with [Rav84].

This book has two goals: to make this material accessible to a general mathematical audience, and to provide algebraic topologists with a coherent and reasonably self-contained account of this material. The nine chapters of the book are directed toward the first goal. The technicalities are suppressed as much as possible, at least in the earlier chapters. The three appendices give descriptions of the tools needed to perform the necessary computations.

In essence almost all of the material of this book can be found in previously published papers. The major exceptions are Chapter 8 (excluding the first section), which hopefully will appear in more detailed form in joint work with Mike Hopkins [HR], and Appendix C, which was recently written up by Jeff Smith [Smi]. In both cases the results were known to their authors by 1986.

This book itself began as a series of twelve lectures given at Northwestern University in 1988, then repeated at the University of Rochester and MSRI (Berkeley) in 1989, at New Mexico State University in 1990, and again at Rochester and Northwestern in 1991. I want to thank all of my listeners for the encouragement that their patience and enthusiasm gave me. Special thanks are due to Sam Gitler and Hal Sadofsky for their careful attention to certain parts of the manuscript.

I am also grateful to all four institutions and to the National Science Foundation for helpful financial support.

D. C. Ravenel
June, 1992

Introduction

In Chapter 1 we will give the elementary definitions in homotopy theory needed to state the main results, the nilpotence theorem (1.4.2) and the periodicity theorem (1.5.4). The latter implies the existence of a global structure in the homotopy groups of many spaces called the chromatic filtration. This is the subject of Chapter 2, which begins with a review of some classical results about homotopy groups.

The nilpotence theorem says that the complex bordism functor reveals a great deal about the homotopy category. This functor and the algebraic category ($C\Gamma$, defined in 3.3.2) in which it takes its values are the subject of Chapters 3 and 4. This discussion is of necessity quite algebraic with the theory of formal group laws playing a major role.

In $C\Gamma$ it is easy to enumerate all the thick subcategories (defined in 3.4.1). The thick subcategory theorem (3.4.3) says that there is a similar enumeration in the homotopy category itself. This result is extremely useful; it means that certain statements about a large class of spaces can be proved by verifying them only for very carefully chosen examples. The thick subcategory theorem is derived from the nilpotence theorem in Chapter 5.

In Chapter 6 we prove the periodicity theorem, using the thick subcategory theorem. First we prove that the set of spaces satisfying the periodicity theorem forms a thick subcategory; this requires some computations in certain noncommutative rings. This thickness statement reduces the proof of the theorem to the construction of a few examples; this requires some modular representation theory due to Jeff Smith.

In Chapter 7 we introduce the concepts of Bousfield localization (7.1.1 and 7.1.3) and Bousfield equivalence (7.2.1). These are useful both for understanding the structure of the homotopy category and for proving the nilpotence theorem. The proof of the nilpotence theorem itself is given in Chapter 9, modulo certain details, for which the reader must consult [DHS88].

There are three appendices which give more technical background for many of the ideas discussed in the text. Appendix A recalls relevant facts known to most homotopy theorists while Appendix B gives more specialized information related to complex bordism theory and BP-theory. Appendix C, which is still more technical, describes some results about representations of the symmetric group due to Jeff Smith [Smi].

The appendices are intended to enable a (sufficiently motivated) nonspecialist to follow the proofs of the text in detail. However, as an introduction to homotopy theory they are very unbalanced. By no means should they be regarded as a substitute for a more thorough study of the subject.

We will now spell out the relation between the conjectures stated in the last section (and listed on the last page) of [Rav84] and the theorems proved here, in the order in which they were stated there. Part (a) of the nilpotence conjecture is the self-map form of the nilpotence theorem, 1.4.2, and part (b) is essentially the smash product form, 5.1.4. Part (c) is the periodicity theorem, 1.5.4, of which the realizabibilty conjecture is an immediate consequence. (This is not quite true since we do not prove that the self-map can be choosen so that its cofibre is a ring spectrum. This has been proved recently by Devinatz [Dev].) The class invariance conjecture is Theorem 7.2.7. The telescope conjecture is stated here as 7.5.5, but is likely to be false in general. The smashing conjecture is the smash product theorem, 7.5.6, and the localization conjecture is Theorem 7.5.2. Finally, the Boolean algebra conjecture, slightly modified to avoid problems with the telescope conjecture, is Theorem 7.2.9.

Two major results proved here that were *not* conjectured in [Rav84] are the thick subcategory theorem (3.4.3) and the chromatic convergence theorem (7.5.7).

Chapter 1

The main theorems

The aim of this chapter is to state the nilpotence and periodicity theorems
(1.4.2 and 1.5.4) with as little technical fussing as possible. Readers familiar
with homotopy theory can skip the first three subsections, which contain
some very elementary definitions.

1.1 Homotopy

A basic problem in homotopy theory is to classify continuous maps up to
homotopy. Two continuous maps from a topological space X to Y are
homotopic if one can be continuously deformed into the other. A more
precise definition is the following.

Definition 1.1.1 *Two continuous maps f_0 and f_1 from X to Y are* **homotopic** *if there is a continuous map (called a homotopy)*

$$X \times [0,1] \xrightarrow{h} Y$$

*such that for $t = 0$ or 1, the restriction of h to $X \times \{t\}$ is f_t. If f_1 is a
constant map, i.e., one that sends all of X to a single point in Y, then
we say that f_0 is* **null homotopic** *and that h is a null homotopy. A
map which is not homotopic to a constant map is* **essential**. *The set of
homotopy classes of maps from X to Y is denoted by $[X, Y]$.*

*For technical reasons it is often convenient to consider maps which send
a specified point $x_0 \in X$ (called the* **base point**) *to a given point $y_0 \in Y$,
and to require that homotopies between such maps send all of $\{x_0\} \times [0,1]$
to y_0. Such maps and homotopies are said to be* **base point preserving**.
*The set of equivalence classes of such maps (under base point preserving
homotopies) is denoted by $[(X, x_0), (Y, y_0)]$*

1

Under mild hypotheses (needed to exclude pathological cases), if X and Y are both path-connected and Y is simply connected, the sets $[X, Y]$ and $[(X, x_0), (Y, y_0)]$ are naturally isomorphic.

In many cases, e.g. when X and Y are compact manifolds or algebraic varieties over the real or complex numbers, this set is countable. In certain cases, such as when Y is a topological group, it has a natural group structure. This is also the case when X is a suspension (1.3.1 and 2.1.2).

In topology two spaces are considered identical if there is a homeomorphism (a continuous map which is one-to-one and onto and which has a continuous inverse) between them. A homotopy theorist is less discriminating than a point set topologist; two spaces are identical in his eyes if they satisfy a much weaker equivalence relation defined as follows.

Definition 1.1.2 *Two spaces X and Y are* **homotopy equivalent** *if there are continuous maps $f: X \to Y$ and $g: Y \to X$ such that gf and fg are homotopic to the identity maps on X and Y. The maps f and g are* **homotopy equivalences**. *A space that is homotopy equivalent to a single point is* **contractible**. *Spaces which are homotopy equivalent have the same* **homotopy type**.

For example, every real vector space is contractible and a solid torus is homotopy equivalent to a circle.

1.2 Functors

In algebraic topology one devises ways to associate various algebraic structures (groups, rings, modules, etc.) with topological spaces and homomorphisms of the appropriate sort with continuous maps.

Definition 1.2.1 *A* **covariant functor** *F from the category of topological spaces T to some algebraic category A (such as that of groups, rings, modules, etc.) is a function which assigns to each space X an object $F(X)$ in A and to each continuous map $f: X \to Y$ a homomorphism $F(f): F(X) \to F(Y)$ in such a way that $F(fg) = F(f)F(g)$ and F sends identity maps to identity homomorphisms. A* **contravariant functor** *G is similar function which reverses the direction of arrows, i.e., $G(f)$ is a homomorphism from $G(Y)$ to $G(X)$ instead of the other way around. In either case a functor is* **homotopy invariant** *if it takes isomorphic values on homotopy equivalent spaces and sends homotopic maps to the same homomorphism.*

Familiar examples of such functors include ordinary homology, which is covariant and cohomology, which is contravariant. Both of these take values in the category of graded abelian groups. Definitions of them can be found in any textbook on algebraic topology. We will describe some less familiar functors which have proved to be extremely useful below.

These functors are typically used to prove that some geometric construction does *not* exist. For example one can show that the 2-sphere S^2 and the torus T^2 (doughnut-shaped surface) are not homeomorphic by computing their homology groups and observing that they are not the same.

Each of these functors has that property that if the continuous map f is null homotopic then the homomorphism $F(f)$ is trivial, but *the converse is rarely true*. Some of the best theorems in the subject concern special situations where it is. One such result is the nilpotence theorem (1.4.2), which is the main subject of the book.

Other results of this type in the past decade concern cases where at least one of the spaces is the classifying space of a finite or compact Lie group. A comprehensive book on this topic has yet to be written. A good starting point in the literature is the J. F. Adams issue of *Topology* (Vol. 31, No. 1, January 1992), specifically [Car92], [DMW92], [JM92], [MP92], and [BF92].

The dream of every homotopy theorist is a solution to the following.

Problem 1.2.2 *Find a functor F from the category of topological spaces to some algebraic category which is reasonably easy to compute and which has the property that $F(f) = 0$ if and only if f is null homotopic.*

We know that this is impossible for several reasons. First, the category of topological spaces is too large. One must limit oneself to a restricted class of spaces in order to exclude many pathological examples which would otherwise make the problem hopeless. Experience has shown that a reasonable class is that of *CW-complexes*. A definition (A.1.1) is given in the Appendix. This class includes all the spaces that one is ever likely to want to study in a geometric way, e.g. all manifolds and algebraic varieties (with or without singularities) over the real or complex numbers. It does not include spaces such as the rational numbers, the p-adic integers or the Cantor set. An old result of Milnor [Mil59] (stated below in Appendix A as A.1.4) asserts that the space of maps from one compact CW-complex to another is homotopy equivalent to a CW-complex. Thus we can include, for example, the space of closed curves on a manifold.

The category of CW-complexes (and spaces homotopy equivalent to them) is a convenient place to do homotopy theory, but in order to have any chance of solving 1.2.2 we must restrict ourselves further by requiring that our complexes be *finite*, which essentially means compact up to homotopy equivalence.

It is convenient to weaken the problem somewhat further. We need another elementary definition from homotopy theory.

1.3   Suspension

Definition 1.3.1 *The* **suspension** *of* X, ΣX *is the space obtained from* $X \times [0,1]$ *by identifying all of* $X \times \{0\}$ *to a single point and all of* $X \times \{1\}$ *to another point. Given a continuous map* $f: X \to Y$, *we define*

$$X \times [0,1] \xrightarrow{\tilde{f}} Y \times [0,1]$$

by $\tilde{f}(x,t) = (f(x),t)$. *This* $\tilde{f}$ *is compatible with the identifications above and gives a map*

$$\Sigma X \xrightarrow{\Sigma f} \Sigma Y.$$

This construction can be iterated and the ith *iterate is denoted by* Σ^i. *If* $\Sigma^i f$ *is null homotopic for some* i *we say that* f *is* **stably null homotopic***; otherwise it is* **stably essential***.*

One can use the suspension to convert $[X,Y]$ *to a graded object* $[X,Y]_*$, *where* $[X,Y]_i = [\Sigma^i X, Y]$. *(We will see below in 2.1.2 that this set has a natural group structure for* $i > 0$.) *It is also useful to consider the group of* **stable homotopy classes of maps***,* $[X,Y]_i^S = \lim_{\to}[\Sigma^{i+j} X, \Sigma^j Y]$.

If X *has a base point* x_0, *we will understand* ΣX *to be the* **reduced suspension***, which is obtained from the suspension defined above by collapsing all of* $\{x_0\} \times [0,1]$ *to (along with* $X \times \{1\}$ *and* $X \times \{0\}$ *) a single point, which is the base point of* ΣX. *(Under mild hypotheses on* X, *the reduced and unreduced suspensions are homotopy equivalent, so we will not distinguish them notationally.)*

Thus ΣX can be thought of as the double cone on X. If S^n (the n-sphere) denotes the space of unit vectors in $\mathbf{R}^{n+1}$, then it is an easy exercise to show that ΣS^n is homeomorphic to S^{n+1}.

Most of the functors we will consider are *homology theories* or, if they are contravariant, *cohomology theories*; the definition will be given below in A.3.3. Ordinary homology and cohomology are examples of such, while homotopy groups (to be defined below in 2.1.1) are not. Classical K-theory is an example of a cohomology theory. Now we will point the properties of such functors that are critical to this discussion.

A homology theory E_* is a functor from the category of topological spaces and homotopy classes of maps to the category of graded abelian groups. This means that for each space X and each integer i, we have an abelian group $E_i(X)$. $E_*(X)$ denotes the collection of these groups for all i. A continuous map $f: X \to Y$ induces a homomorphism

$$E_i(X) \xrightarrow{E_i(f)} E_i(Y)$$

which depends only on the homotopy class of f.

In particular one has a canonical homomorphism

$$E_*(X) \xrightarrow{\epsilon} E_*(\text{pt.}),$$

called the *augmentation map*, induced by the constant map on X. Its kernel, denoted by $\overline{E}_*(X)$, is called the *reduced homology* of X, while $E_*(X)$ is sometimes called the *unreduced homology* of X.

Note that the augmentation is the projection onto a direct summand because one always has maps

$$\text{pt.} \longrightarrow X \longrightarrow \text{pt.}$$

whose composite is the identity. $E_*(\text{pt.})$ is nontrivial as long as E_* is not identically zero. A reduced homology theory vanishes on every contractible space.

One of the defining axioms of a homology theory (see A.3.3) implies that there is a natural isomorphism

$$\overline{E}_i(X) \xrightarrow{\sigma} \overline{E}_{i+1}(\Sigma X) \qquad (1.3.2)$$

A *multiplicative homology theory* is one equipped with a ring structure on $E_*(\text{pt.})$ (which is called the *coefficient ring* and usually denoted simply by E_*), over which $E_*(X)$ has a functorial module structure.

Problem 1.3.3 *Find a reduced homology theory $\overline{E}_*$ on the category of finite CW-complexes which is reasonably easy to compute and which has the property that $F(f) = 0$ if and only if $\Sigma^i f$ is null homotopic for some i.*

In this case there is a long standing conjecture of Freyd [Fre66, §9], known as the generating hypothesis, which says that stable homotopy (to be defined in 2.2.3) is such a homology theory. A partial solution to the problem, that is very much in the spirit of this book, is given by Devinatz in [Dev90].

(The generating hypothesis was arrived in the following way. The stable homotopy category **FH** of finite complexes is additive, that is the set of morphisms between any two objects has a natural abelian group structure. Freyd gives a construction for embedding any additive category into an abelian category, i.e., one with kernels and cokernels. It is known that any abelian category is equivalent to a category of modules over some ring. This raises the question of identifying the ring thus associated with **FH**. It is natural to guess that it is π_*^S, the stable homotopy groups of spheres. This statement is equivalent to the generating hypothesis.)

Even if the generating hypothesis were known to be true, it would not be a satisfactory solution to 1.3.3 because stable homotopy groups are anything but easy to compute.

1.4   Self-maps and the nilpotence theorem

Now suppose that the map we want to study has the form

$$\Sigma^d X \xrightarrow{f} X$$

for some $d \geq 0$. Then we can iterate it up to suspension by considering the composites

$$\cdots \Sigma^{3d} X \xrightarrow{\Sigma^{2d} f} \Sigma^{2d} X \xrightarrow{\Sigma^d f} \Sigma^d X \xrightarrow{f} X$$

For brevity we denote these composite maps to X by f, f^2, f^3, etc.

Definition 1.4.1 *A map* $f \colon \Sigma^d X \to X$ *is a* **self-map** *of X. It is* **nilpotent** *if some suspension of* f^t *for some* $t > 0$ *is null homotopic. Otherwise we say that f is* **periodic**.

If we apply a reduced homology theory $\overline{E}_*$ to a self-map f, by 1.3.2 we get an endomorphism of $\overline{E}_*(X)$ that raises the grading by d.

Now we can state the nilpotence theorem of Devinatz-Hopkins-Smith [DHS88].

Theorem 1.4.2 (Nilpotence theorem, self-map form) *There is a homology theory MU_* such that a self-map f of a finite CW-complex X is stably nilpotent if and only if some iterate of $\overline{MU}_*(f)$ is trivial.*

Actually this is only one of three equivalent forms of the nilpotence theorem; we will state the other two below (5.1.4 and 9.0.1).

The functor MU_*, known as complex bordism theory , takes values in the category of graded modules over a certain graded ring L, which is isomorphic to $MU_*(\text{pt.})$. These modules come equipped with an action by a certain infinite group Γ, which also acts on L. The ring L and the group Γ are closely related to the theory of formal group laws. $MU_*(X)$ was originally defined in terms of maps from certain manifolds to X, but this definition sheds little light on its algebraic structure. It is the algebra rather than the geometry which is central to our discussion. We will discuss this in more detail in Section 3 and more background can be found in [Rav86, Chapter 4]. In practice it is not difficult to compute, although there are still plenty of interesting spaces for which it is still unknown.

1.5   Morava K-theories and the periodicity theorem

We can also say something about periodic self-maps.

Before doing so we must discuss localization at a prime p. In algebra one does this by tensoring everything in sight by $\mathbf{Z}_{(p)}$, the integers localized at the prime p; it is the subring of the rationals consisting of fractions with denominator prime to p. If A is a finite abelian group, then $A \otimes \mathbf{Z}_{(p)}$ is the p-component of A. $\mathbf{Z}_{(p)}$ is flat as a module over the integers $\mathbf{Z}$; this means that tensoring with it preserves exact sequences.

There is an analogous procedure in homotopy theory. The definitive reference is [BK72]; a less formal account can be found in [Ada75]. For each CW-complex X there is a unique $X_{(p)}$ with the property that for any homology theory E_*, $\overline{E}_*(X_{(p)}) \cong \overline{E}_*(X) \otimes \mathbf{Z}_{(p)}$. We call $X_{(p)}$ the p-localization of X. If X is finite we say $X_{(p)}$ is a p-local finite CW-complex.

Proposition 1.5.1 *Suppose X is a simply connected CW-complex such that $\overline{H}_*(X)$ consists entirely of torsion.*

(i) If this torsion is prime to p then $X_{(p)}$ is contractible.

(ii) If it is all p-torsion then X is p-local, i.e., $X_{(p)}$ is equivalent to X. (In this case we say that X is a p-torsion complex.)

(iii) In general X is homotopy equivalent to the one-point union of its p-localizations for all the primes p in this torsion.

If X is as above, then its p-localization will be nontrivial only for finitely many primes p. The cartesian product of any two of them will be the same as the one-point union. The smash product (defined below in 5.1.2)

$$X_{(p)} \wedge X_{(q)}$$

is contractible for distinct primes p and q.

The most interesting periodic self-maps occur when X is a finite p-torsion complex. In these cases it is convenient to replace MU_* by the Morava K-theories. These were invented by Jack Morava, but he never published an account of them. Most of the following result is proved in [JW75]; a proof of (v) can be found in [Rav84].

Proposition 1.5.2 *For each prime p there is a sequence of homology theories $K(n)_*$ for $n \geq 0$ with the following properties. (We follow the standard practice of omitting p from the notation.)*

(i) $K(0)_*(X) = H_*(X; \mathbf{Q})$ and $\overline{K(0)}_*(X) = 0$ when $\overline{H}_*(X)$ is all torsion.

(ii) $K(1)_*(X)$ is one of $p-1$ isomorphic summands of mod p complex K-theory.

(iii) $K(0)_*(\mathrm{pt.}) = \mathbf{Q}$ and for $n > 0$, $K(n)_*(\mathrm{pt.}) = \mathbf{Z}/(p)[v_n, v_n^{-1}]$ where the dimension of v_n is $2p^n - 2$. This ring is a graded field in the sense that every graded module over it is free. $K(n)_*(X)$ is a module over $K(n)_*(\mathrm{pt.})$.

(iv) There is a Künneth isomorphism

$$K(n)_*(X \times Y) \cong K(n)_*(X) \otimes_{K(n)_*(pt.)} K(n)_*(Y).$$

(v) Let X be a p-local finite CW-complex. If $\overline{K(n)}_*(X)$ vanishes , then so does $\overline{K(n-1)}_*(X)$.
(vi) If X as above is not contractible then

$$\overline{K(n)}_*(X) = K(n)_*(pt.) \otimes \overline{H}_*(X; \mathbf{Z}/(p))$$

for n sufficiently large. In particular it is nontrivial if X is not contractible.

Definition 1.5.3 *A p-local finite complex X has* **type** *n if n is the smallest integer such that $\overline{K(n)}_*(X)$ is nontrivial. If X is contractible it has type ∞.*

Because of the Künneth isomorphism, $K(n)_*(X)$ is easier to compute than $MU_*(X)$. Again there are still many interesting spaces for which this has not been done. See [RW80] and [HKR]. A corollary of the nilpotence theorem (1.4.2) says that the Morava K-theories, along with ordinary homology with coefficients in a field, are essentially the only homology theories with Künneth isomorphisms.

The Morava K-theories for $n > 0$ have another property which we will say more about below. Suppose we ignore the grading on $K(n)_*(X)$ and consider the tensor product

$$K(n)_*(X) \otimes_{K(n)_*(pt.)} \mathbf{F}_{p^n}$$

where $\mathbf{F}_{p^n}$ denotes the field with p^n elements, which is regarded as a module over $K(n)_*(pt.)$ by sending v_n to 1. Then this $\mathbf{F}_{p^n}$-vector space is acted upon by a certain p-adic Lie group S_n (not to be confused with the n-sphere S^n) which is contained in a certain p-adic division algebra.

The Morava K-theories are especially useful for detecting periodic self-maps. This is the subject of the second major result of this book, the periodicity theorem of Hopkins-Smith [HS]. The proof is outlined in [Hop87] and in Chapter 6.

Theorem 1.5.4 (Periodicity theorem) *Let X and Y be p-local finite CW-complexes of type n (1.5.3) for n finite.*
(i) There is a self-map $f: \Sigma^{d+i}X \to \Sigma^i X$ for some $i \geq 0$ such that $K(n)_*(f)$ is an isomorphism and $K(m)_*(f)$ is trivial for $m > n$. (We will refer to such a map as a v_n-map.) When $n = 0$ then $d = 0$, and when $n > 0$ then d is a multiple of $2p^n - 2$.
(ii) Suppose $h: X \to Y$ is a continuous map. Assume that both have already been suspended enough times to be the target of a v_n-map. Let

$g: \Sigma^e Y \to Y$ *be a self-map as in (i). Then there are positive integers i and j with $di = ej$ such that the following diagram commutes up to homotopy.*

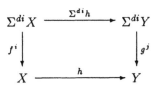

(The integers i and j can be chosen independently of the map h.)

The map h in (ii) could be the identity map, which shows that f is *assymptotically unique* in the following sense. Suppose g is another such periodic self-map. Then there are positive integers i and j such that f^i is homotopic to g^j. If X is a suspension of Y and f is a suspension of g, this shows that f is assymptotically central in that any map h commutes with some iterate of f.

Chapter 2

Homotopy groups and the chromatic filtration

In this section we will describe the homotopy groups of spheres, which make up one of the messiest but most fundamental objects in algebraic topology. First we must define them.

2.1 The definition of homotopy groups

The following definition is originally due to Čech [Cec32]. Homotopy groups were first studied sytematically by Witold Hurewicz in [Hur35] and [Hur36].

Definition 2.1.1 *The nth **homotopy group of** X, $\pi_n(X)$ is the set of homotopy classes of maps from the n-sphere S^n (the space of unit vectors in $\mathbf{R}^{n+1}$) to X which send a fixed point in S^n (called the base point) to a fixed point in X. (If X is not path-connected, then we must specify in which component its base point x_0 is chosen to lie. In this case the group is denoted by $\pi_n(X, x_0)$.) $\pi_1(X)$ is the **fundamental group** of X.*

We define a group structure on $\pi_n(X)$ as follows. Consider the pinch map

$$S^n \xrightarrow{\quad \text{pinch} \quad} S^n \vee S^n$$

obtained by collapsing the equator in the source to a single point. Here $X \vee Y$ denotes the one-point union of X and Y, i.e., the union obtained by identifying the base point in X with the one in Y. We assume that the base point in the source S^n has been chosen to lie on the equator, so that the map above is base point preserving.

11

Now let $\alpha, \beta \in \pi_n(X)$ be represented by maps $f, g: S^n \to X$. Define $\alpha \cdot \beta \in \pi_n(X)$ to be the class of the composite

$$S^n \xrightarrow{\quad \text{pinch} \quad} S^n \vee S^n \xrightarrow{\quad f \vee g \quad} X.$$

The inverse α^{-1} is obtained by composing f with a base point preserving reflection map on S^n.

It is easy to verify that this group structure is well defined and that it is abelian for $n > 1$.

It is easy to construct a space whose π_1 is any given finitely presented group. This means that certain classification problems in homotopy theory contain problems in group theory that are known to be unsolvable.

Remark 2.1.2 *In a similar way one can define a group structure on the set of base point preserving maps from ΣX to Y for any space X (not just $X = S^{n-1}$ as above) and show that it is abelian whenever X is a suspension, i.e., whenever the source of the maps is a double suspension.*

These groups are easy to define but, unless one is very lucky, quite difficult to compute. Of particular interest are the homotopy groups of the spheres themselves. These have been the subject of a great deal of effort by many algebraic topologists who have developed an arsenal of techniques for calculating them. Many references and details can be found in [Rav86]. We will not discuss any of these methods here, but we will describe a general approach to the problem suggested by the nilpotence and periodicity theorems known as the chromatic filtration.

2.2   Classical theorems

First we need to recall some classical theorems on the subject.

Theorem 2.2.1 (Hurewicz theorem, 1935) *The groups $\pi_n(S^m)$ are trivial for $n < m$, and $\pi_n(S^n) \cong \mathbf{Z}$; this group is generated by the homotopy class of the identity map.*

The next result is due to Hans Freudenthal [Fre37].

Theorem 2.2.2 (Freudenthal suspension theorem, 1937) *The suspension homomorphism (see 1.3.1)*

$$\sigma: \pi_{n+k}(S^n) \to \pi_{n+k+1}(S^{n+1})$$

is an isomorphism for $k < n - 1$. The same is true if we replace S^n by any $(n-1)$-connected space X, i.e., any space X with $\pi_i(X) \cong 0$ for $i < n$.

This means that $\pi_{n+k}(\Sigma^n X)$ depends only on k if $n > k+1$.

Definition 2.2.3 *The* k^{th} **stable homotopy group of** X, $\pi_k^S(X)$, *is*

$$\pi_{n+k}(\Sigma^n X) \text{ for } n > k+1.$$

In particular $\pi_k^S(S^0) = \pi_{n+k}(S^n)$, *for* n *large, is called the* **stable** k**-stem** *and will be abbreviated by* π_k^S.

The stable homotopy groups of spheres are easier to compute than the unstable ones. They are finite for $k > 0$. The p-component of π_k^S is known for $p = 2$ for $k < 60$ and for p odd for $k < 2p^3(p-1)$. Tables for $p = 2, 3$ and 5 can be found in [Rav86]. Empirically we find that $\log_p |(\pi_k^S)_{(p)}|$ grows linearly with k.

The next result is due to Serre [Ser53] and gives a complete description of $\pi_*(S^n)$ mod torsion.

Theorem 2.2.4 (Serre finiteness theorem, 1953) *The homotopy groups of spheres are finite abelian except in the following cases:*

$$\pi_n(S^n) \cong \mathbf{Z} \quad \text{and}$$
$$\pi_{4m-1}(S^{2m}) \cong \mathbf{Z} \oplus F_m$$

where F_m *is finite abelian.*

Before stating the next result we need to observe that π_*^S is a graded ring. If $\alpha \in \pi_k^S$ and $\beta \in \pi_\ell^S$ are represented by maps $f: S^{n+k} \to S^n$ and $g: S^{n+\ell} \to S^n$, then $\alpha\beta \in \pi_{k+\ell}^S$ is represented by the composite

$$S^{n+k+\ell} \xrightarrow{\quad \Sigma^k g \quad} S^{n+k} \xrightarrow{\quad f \quad} S^n.$$

This product is commutative up to the usual sign in algebraic topology, i.e., $\beta\alpha = (-1)^{k+\ell}\alpha\beta$.

The following was proved in [Nis73].

Theorem 2.2.5 (Nishida's theorem, 1973) *Each element in* π_k^S *for* $k > 0$ *is nilpotent, i.e., some power of it is zero.*

This is the special case of the nilpotence theorem for $X = S^n$. It also shows that π_*^S as a ring is very bad; it has no prime ideals other than (p). It would not be a good idea to try to describe it in terms of generators and relations. We will outline another approach to it at the end of this section.

The following result was proved in [CMN79]

Theorem 2.2.6 (Cohen-Moore-Neisendorfer theorem, 1979) *For* p *odd and* $k > 0$, *the exponent of* $\pi_{2n+1+k}(S^{2n+1})_{(p)}$ *is* p^n, *i.e., there are no elements of order* p^{n+1}.

2.3   Cofibres

By the early 1970's several examples of periodic maps had been discovered and used to construct infinite families of elements in the stable homotopy groups of spheres. Before we can describe them we need another elementary definition from homotopy theory.

Definition 2.3.1 *Let $f: X \to Y$ be a continuous map. Its* **mapping cone,** *or* **cofibre,** C_f, *is the space obtained from the disjoint union of $X \times [0,1]$ and Y by identifying all of $X \times \{0\}$ to a single point and $(x,1) \in X \times [0,1]$ with $f(x) \in Y$.*

If X and Y have base points x_0 and y_0 respectively with $f(x_0) = y_0$, then we define C_f to be as above but with all of $\{x_0\} \times [0,1]$ collapsed to a single point, which is defined to be the base point of C_f. (This C_f is homotopy equivalent to the one defined above.)

In either case, Y is a subspace of C_f, and the evident inclusion map will be denoted by i.

The following result is an elementary exercise.

Proposition 2.3.2 *Let $i: Y \to C_f$ be the map given by 2.3.1. Then C_i is homotopy equivalent to ΣX.*

Definition 2.3.3 *A* **cofibre sequence** *is a sequence of spaces and maps of the form*

$$X \xrightarrow{f} Y \xrightarrow{i} C_f \xrightarrow{j} \Sigma X \xrightarrow{\Sigma f} \Sigma Y \longrightarrow \cdots$$

in which each space to the right of Y is the mapping cone of the map preceding the map to it, and each map to the right of f is the canonical inclusion of a map's target into its mapping cone, as in 2.3.1.

If one has a homotopy commutative diagram

$$
\begin{array}{ccccc}
X & \xrightarrow{\;\;f\;\;} & Y & \xrightarrow{\;\;i\;\;} & C_f \\
{\scriptstyle g_1}\big\uparrow & & {\scriptstyle g_2}\big\uparrow & & \big\vdots \\
X' & \xrightarrow{\;\;f'\;\;} & Y' & \xrightarrow{\;\;i'\;\;} & C_{f'}
\end{array}
$$

then the missing map always exists, although it is not unique up to homotopy. Special care must be taken if f' is a suspension of f. Then the diagram extends to

$$
\begin{array}{ccccccc}
X & \xrightarrow{\;\;f\;\;} & Y & \xrightarrow{\;\;i\;\;} & C_f & \xrightarrow{\;\;j\;\;} & \Sigma X \\
{\scriptstyle g_1}\big\uparrow & & {\scriptstyle g_2}\big\uparrow & & {\scriptstyle g_3}\big\uparrow & & {\scriptstyle \Sigma g_1}\big\uparrow \\
\Sigma^d X & \xrightarrow{\;\Sigma^d f\;} & \Sigma^d Y & \xrightarrow{\;\Sigma^d i\;} & \Sigma^d C_f & \xrightarrow{(-1)^d \Sigma^d j} & \Sigma^{d+1} X
\end{array}
$$

and the sign in the suspension of j is unavoidable.

Now suppose $g: Y \to Z$ is continuous and that gf is null homotopic. Then g can be extended to a map $\tilde{g}: C_f \to Z$, i.e., there exists a $\tilde{g}$ whose restriction to Y (which can be thought of as a subspace of C_f) is g. More explicitly, suppose $h: X \times [0,1] \to Z$ is a null homotopy of gf, i.e., a map whose restriction to $X \times \{0\}$ is constant and whose restriction to $X \times \{1\}$ is gf. Combining h and g we have a map to Z from the union of $X \times [0,1]$ with Y which is compatible with the identifications of 2.3.1. Hence we can use h and g to define $\tilde{g}$.

Note that $\tilde{g}$ depends on the homotopy h; a different h can lead to a different (up to homotopy) $\tilde{g}$. The precise nature of this ambiguity is clarified by the following result, which describes two of the fundamental long exact sequences in homotopy theory.

Proposition 2.3.4 *Let X and Y be path connected CW-complexes.*

(i) For any space Z the cofibre sequence of 2.3.3 induces a long exact sequence

$$[X, Z] \xleftarrow{f^*} [Y, Z] \xleftarrow{i^*} [C_f, Z] \xleftarrow{j^*} [\Sigma X, Z] \xleftarrow{\Sigma f^*} [\Sigma Y, Z] \longleftarrow \cdots$$

Note that each set to the right of $[C_f, Z]$ is a group, so exactness is defined in the usual way, but the first three sets need not have group structures. However each of them has a distinguished element, namely the homotopy class of the constant map. Exactness in this case means that the image of one map is the preimage of the constant element under the next map.

(ii) Let E_* be a homology theory. Then there is a long exact sequence

$$\cdots \xrightarrow{j_*} \overline{E}_m(X) \xrightarrow{f_*} \overline{E}_m(Y) \xrightarrow{i_*} \overline{E}_m(C_f) \xrightarrow{j_*} \overline{E}_{m-1}(X) \xrightarrow{f_*} \cdots$$

(iii) Suppose X and Y are each $(k-1)$-connected (i.e., their homotopy groups vanish below dimension k) and let W be a finite CW-complex (see A.1.1) which is a double suspension with top cell in dimension less than $2k - 1$. Then there is a long exact sequence of abelian groups

$$[W, X] \xrightarrow{f_*} [W, Y] \xrightarrow{i_*} [W, C_f] \xrightarrow{j_*} [W, \Sigma X] \xrightarrow{\Sigma f_*} [W, \Sigma Y] \longrightarrow \cdots.$$

This sequence will terminate at the point where the connectivity of the target exceeds the dimension of W.

Corollary 2.3.5 *Suppose X as in the periodicity theorem (1.5.4) has type n. Then the cofibre of the map given by 1.5.4 has type $n + 1$.*

Proof. Assume that X has been suspended enough times to be the target of a v_n-map f and let W be its cofibre. We will study the long exact sequence

$$\cdots \xrightarrow{j_*} \overline{K(m)}_t(\Sigma^d X) \xrightarrow{f_*} \overline{K(m)}_t(X) \xrightarrow{i_*} \overline{K(m)}_t(W) \xrightarrow{j_*} \overline{K(m)}_{t-1}(\Sigma^d X) \xrightarrow{f_*} \cdots$$

for various m.

For $m < n$, $\overline{K(m)}_*(X) = 0$, so $\overline{K(m)}_*(W) = 0$. For $m = n$, f_* is an isomorphism, so again $\overline{K(m)}_*(W) = 0$. For $m > n$, $f_* = 0$ and $\overline{K(m)}_*(X) \neq 0$ by 1.5.2(v). It follows that

$$\overline{K(m)}_*(W) \cong \overline{K(m)}_*(X) \oplus \overline{K(m)}_*(\Sigma^{d+1}X),$$

so W has type $n + 1$.   ∎

2.4   Motivating examples

The following examples of periodic maps led us to conjecture the nilpotence and periodicity theorems.

Example 2.4.1 (The earliest known periodic maps) *(i) Regard S^1 as the unit circle in the complex numbers* **C**. *The degree p map on S^1 is the one which sends z to z^p. This map is periodic in the sense of 1.4.1, as is each of its suspensions. In this case n (as in the periodicity theorem) is zero.*

(ii) Let $V(0)_k$ (known as the mod p *Moore space) be the cofibre of the degree p map on S^k. Adams [Ada66a] and Toda [Tod60] showed that for sufficiently large k there is a periodic map*

$$\Sigma^q V(0)_k \xrightarrow{\ \alpha\ } V(0)_k$$

where q is 8 when $p = 2$ and $2p - 2$ for p odd. In this case the n of 1.5.4 is one. The induced map in $K(1)_*(V(0)_k)$ is multiplication by v_1 when p is odd and by v_1^4 for $p = 2$.

For $p = 2$ there is no self-map inducing multiplication by a smaller power of v_1. One could replace the mod 2 Moore space by the mod 16 Moore space and still have a map α as above.

(iii) For $p \geq 5$, let $V(1)_k$ denote the cofibre of the map in (ii). Larry Smith [Smi71] and H. Toda [Tod71] showed that for sufficiently large k there is a periodic map

$$\Sigma^{2p^2-2} V(1)_k \xrightarrow{\ \beta\ } V(1)_k$$

which induces multiplication by v_2 in $K(2)$-theory.

(iv) For $p \geq 7$, let $V(2)_k$ denote the cofibre of the map in (iii). Smith and Toda showed that for sufficiently large k there is a periodic map

$$\Sigma^{2p^3-2} V(2)_k \xrightarrow{\ \gamma\ } V(2)_k$$

which induces multiplication by v_3 in $K(3)$-theory. We denote its cofibre by $V(3)_k$.

These results were not originally stated in terms of Morava K-theory, but in terms of complex K-theory in the case of (ii) and complex bordism in the case of (iii) and (iv). Attempts to find a self-map on $V(3)$ inducing multiplication by v_4 have been unsuccessful. The Periodicity Theorem guarantees that there is a map inducing multiplication by some power of v_4, but gives no upper bound on the exponent. References to some other explicit examples of periodic maps can be found in [Rav86, Chapter 5].

Each of the maps in 2.4.1 led to an infinite family (which we also call *periodic*) of elements in the stable homotopy groups of spheres as follows.

Example 2.4.2 (Periodic families from periodic maps) *(i) We can iterate the degree p map of 2.4.1(i) and get multiples of the identity map on S^k by powers of p, all of which are essential.*

(ii) With the map α of 2.4.1(ii) we can form the following composite.

$$S^{k+qt} \xrightarrow{i_1} \Sigma^{qt} V(0)_k \xrightarrow{\alpha^t} V(0)_k \xrightarrow{j_1} S^{k+1}$$

where $i_1: S^k \to V(0)_k$ and $j_1: V(0)_k \to S^{k+1}$ are maps in the cofibre sequence associated with the degree p map. (We are using the same notation for a map and each of its suspensions) This composite was shown by Adams [Ada66a] to be essential for all $t > 0$. The resulting element in π^S_{qt-1} is denoted by α_t for p odd and by α_{4t} for $p = 2$.

(iii) Let $i_2: V(0)_k \to V(1)_k$ and $j_2: V(1)_k \to \Sigma^{q+1} V(0)_k$ denote the maps in the cofibre sequence associated with α. Using the map β of 2.4.1(iii) for $p \geq 5$ we have the composite

$$S^{k+2(p^2-1)t} \xrightarrow{i_2 i_1} \Sigma^{k+2(p^2-1)t} V(1)_k \xrightarrow{\beta^t} V(1)_k \xrightarrow{j_1 j_2} S^{k+2p}$$

which is denoted by $\beta_t \in \pi_{2(p^2-1)t-2p}$. Smith [Smi71] showed it is essential for all $t > 0$.

(iv) For $p \geq 7$ there is a similarly defined composite

$$S^{k+2(p^3-1)t} \longrightarrow \Sigma^{k+2(p^3-1)t} V(2)_k \xrightarrow{\gamma^t} V(2)_k \longrightarrow S^{k+(p+2)q+3}$$

which is denoted by γ_t. It was shown to be nontrivial for all $t > 0$ in [MRW77].

In general a periodic map on a finite CW-complex leads to a periodic family of elements in π^S_*, although the procedure is not always as simple as in the above examples. Each of them has the following features. We have a CW-complex (defined below in A.1.1) X of type n with bottom cell in dimension k and top cell in some higher dimension, say $k + e$. Thus we have an inclusion map $i_0 : S^k \to X$ and a pinch map $j_0 : X \to S^{k+e}$. Furthermore the composite

$$S^{k+td} \xrightarrow{i_0} \Sigma^{td}X \xrightarrow{f^t} X \xrightarrow{j_0} S^{k+e} \qquad (2.4.3)$$

is essential for each $t > 0$, giving us a nontrivial element in π^S_{td-e}. This fact does *not* follow from the nontriviality of f^t; in each case a separate argument (very difficult in the case of the γ_t) is required.

If the composite (2.4.3) is null, we can still get a nontrivial element in $\pi^S_{td-\epsilon}$ (for some ϵ between e and $-e$) as follows. At this point we need to be in the stable range, i.e., we need $k > td + e$, so we can use A.1.6. This can be accomplished by suspending everything in sight enough times.

For $k \leq r \leq s \leq k + e$, X^s_r will denote the cofibre of the inclusion map $X^{r-1} \to X^s$. In particular, $X^{k+e}_k = X$ and X^s_s is a wedge of s-spheres, one for each s-cell in X. We will use the letter i to denote any inclusion map $X^s_r \to X^{s'}_r$ with $s' > s$, and the letter j to denote any pinch map $X^s_r \to X^s_{r'}$ with $s \geq r' > r$.

Now let $f_e = f^t$ and now consider the diagram

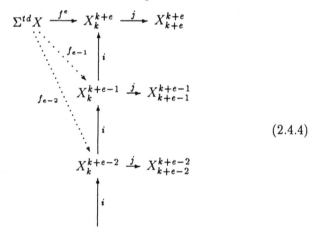

$$(2.4.4)$$

If the composite jf_e is null, then by 2.3.4(iii) there is a map f_{e-1} with $if_{e-1} = f_e$. Similarly if jf_{e-1} is null then there is a map f_{e-2} with $if_{e-2} = f_{e-1}$. We proceed in this way until the composite

$$\Sigma^{td}X \xrightarrow{f_{e_1}} X^{k+e_1}_k \xrightarrow{j} X^{k+e_1}_{k+e_1}$$

is essential. This must be the case for some e_1 between 0 and e, because if all of those composites were null, then 2.3.4(iii) would imply that f^t is null.

Now let $g_0 = jf_{e_1}$ and consider the diagram

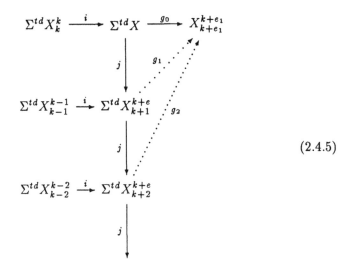

$$(2.4.5)$$

This time we use 2.3.4(i) instead of 2.3.4(iii). It says that if $g_0 i$ is null then there is a map g_1 with $g_1 j = g_0$. Similarly if $g_1 i$ is null there is a map g_2 with $g_1 = g_2 j$. The composites $g_m i$ for $0 \leq m \leq e$ cannot all be null because g_0 is essential. Let e_2 be the integer between 0 and e such that $g_{e_2} i$ is essential.

Summing up, we have a diagram

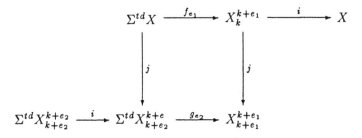

where $i f_{e_1} = f^t$. The source and target of $g_{e_2} i$ are both wedges of spheres, so this is the promised stable homotopy element. Its dimension is $td + e_2 - e_1$ with $0 \leq e_1, e_2 \leq e$.

The simplest possible outcome of this procedure is the case $e_1 = e$ and $e_2 = 0$; this occurs in each of the examples in 2.4.2. In any other outcome, the construction is riddled with indeterminacy, because the maps f_{e_1} and g_{e_2} are not unique.

In any case the outcome may vary with the exponent t. In every example that we have been able to analyze, the behavior is as follows. With a finite number of exceptions (i.e., for t sufficiently large), the outcome depends only on the congruence class of t modulo some power of the prime p.

2.5   The chromatic filtration

These examples led us to ask if every element in the stable homotopy groups of spheres is part of such a family. In [MRW77] we explored an algebraic analog of this question. The Adams-Novikov spectral sequence (A.6.3) is a device for computing π_*^S and its E_2-term was shown there to have such an organization using a device called the chromatic spectral sequence(see B.8), which is also described in [Rav86, Chapter 5]. In [Rav84] we explored the question of making this algebraic structure more geometric. It was clear that the periodicity theorem would be essential to this program, and that the former would be false if there were a counter example to the nilpotence theorem. Now that the nilpotence and periodicity theorems have been proved, we can proceed directly to the geometric construction that we were looking for in [Rav84] without dwelling on the details of the chromatic spectral sequence.

Suppose Y is a p-local complex and $y \in \pi_k(Y)$ is represented by a map $g: S^k \to Y$. If all suspensions of y have infinite order, then it has a nontrivial image in $\pi_k^S(Y) \otimes \mathbf{Q}$. In the case where Y is a sphere, this group can be read off from 2.2.4. In general this group is easy to compute since it is known to isomorphic to $\overline{H}_*(X; \mathbf{Q})$.

On the other hand, if some suspension of y has order p^i, then it factors through the cofibre of the map of degree p^i on the corresponding suspension of S^k, which we denote here by $W(1)$. For the sake of simplicity we will ignore suspensions in the rest of this discussion. The map from $W(1)$ to Y will be denoted by g_1.

The complex $W(1)$ has type 1 and therefore a periodic self-map

$$f_1: \Sigma^{d_1} W(1) \to W(1)$$

which induces a $K(1)_*$-equivalence. Now we can ask whether g_1 becomes null homotopic when composed with some iterate of f_1 or not. If all such composites are stably essential then g_1 has a nontrivial image in the direct limit obtained by taking homotopy classes of maps form the inverse system

$$W(1) \xleftarrow{f_1} \Sigma^{d_1} W(1) \xleftarrow{f_1} \Sigma^{2d_1} W(1) \longleftarrow \cdots$$

which gives a direct system of groups

$$[W(1), Y]_*^S \xrightarrow{f_1^*} [\Sigma^{d_1} W(1), Y]_*^S \xrightarrow{f_1^*} [\Sigma^{2d_1} W(1), Y]_*^S \xrightarrow{f_1^*} \cdots, \qquad (2.5.1)$$

which we denote by $v_1^{-1}[W(1), Y]_*^S$. Note that the second part of the periodicity theorem implies that this limit is independent of the choice of f_1.

This group was determined in the case when Y is a sphere for the prime 2 by Mahowald in [Mah81] and for odd primes by Miller in [Mil81]. More precise calculations not requiring any suspensions of the spaces in question in the case when Y is an odd-dimensional sphere were done for $p = 2$ by Mahowald in [Mah82] and for p odd by Thompson in [Tho90]. In general it appears to be an accessible problem. For more details, see [Ben92], [BD], [BDM], [Dav91] and [DM].

There is a definition of $v_1^{-1}\pi_*(Y)$ which is independent of the exponent i. We have an inverse system of cofibre sequences

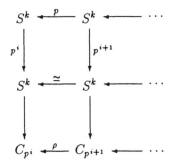

which induces a direct system of long exact sequences

$$\cdots \longrightarrow [C_{p^i}, Y] \longrightarrow \pi_k(Y) \xrightarrow{p^i} \pi_k(Y) \longrightarrow \cdots$$

$$\cdots \longrightarrow [C_{p^{i+1}}, Y] \longrightarrow \pi_k(Y) \xrightarrow{p^{i+1}} \pi_k(Y) \longrightarrow \cdots$$

The limit of these is a long exact sequence of the form

$$\cdots \longrightarrow \pi_k(Y) \longrightarrow p^{-1}\pi_k(Y) \longrightarrow \pi_k(Y)/(p^\infty) \longrightarrow \cdots$$

where

$$\pi_k(Y)/(p^\infty) = \varinjlim [\Sigma^{-1} C_{p^i}, Y].$$

Note that since Y is p-local,

$$p^{-1}\pi_k(Y) = \pi_k(Y) \otimes \mathbf{Q}.$$

We can define $v_1^{-1}\pi_*(Y)/(p^\infty)$ by using some more detailed information about v_1-maps on the Moore spaces C_{p^i}. For suffciently large k (independent of i), there are v_1-maps

$$\Sigma^{2p^{i-1}(p-1)}C_{p^i} \xrightarrow{f_{1,i}} C_{p^i}$$

such that the following diagram commutes.

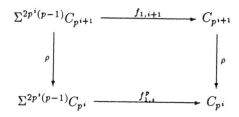

This means we can define the groups $v_1^{-1}[C_{p^i}, Y]_*$ compatibly for various i, and their direct limit is $v_1^{-1}\pi_*(Y)/(p^\infty)$. More details of this construction can be found in [DM].

Returning to (2.5.1), suppose that some power of f_1 annihilates g_1, i.e., some composite of the form $g_1 f_1^i$ is stably null homotopic. In this case, let $W(2)$ be the cofibre of f_1^i and let g_2 be an extension of g_1 to $W(2)$.

Then $W(2)$ has type 2 and therefore it admits a periodic self-map

$$\Sigma^{d_2}W(2) \xrightarrow{f_2} W(2)$$

which is detected by $K(2)_*$. This leads us to consider the group

$$v_2^{-1}[W(2), Y]_*^S.$$

This group is not yet known for any Y. There is some machinery (see 7.5 and B.8) available for computing what was thought to be a close algebraic approximation. The relation of this approximation to the actual group in question was the subject of the telescope conjecture (7.5.5), which has recently been disproved by the author. (It is known to be true in the v_1 case.) The algebraic computation in the case where Y is a sphere, $W(1)$ is a mod p Moore space and $p \geq 5$ has been done by Shimomura and Tamura in [Shi86] and [ST86].

Summing up, we have a diagram

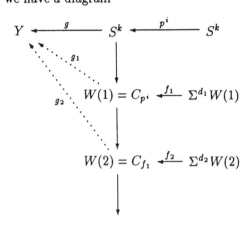

One could continue this process indefinitely. At the nth stage one has an extension g_n of the original map g to a complex $W(n)$ of type n which has a periodic self-map f_n. Then one asks if g_n is annihilated stably by some iterate of f_n. If the answer is no, then the process stops and g_n has a nontrivial image in the group $v_n^{-1}[W(n), Y]_*^S$. On the other hand, if g_n is annihilated by a power of f_n then we can move on to the $(n+1)^{\text{th}}$ stage.

In view of this we make the following definitions.

Definition 2.5.2 *If an element $y \in \pi_*^S(Y)$ extends to a complex $W(n)$ of type n as above, then y is v_{n-1}-torsion. If in addition y does not extend to a complex of type $n + 1$, it is v_n-periodic. The **chromatic filtration** of $\pi_*^S(Y)$ is the decreasing family of subgroups consisting of the v_n-torsion elements for various $n \geq 0$.*

We use the word 'chromatic' here for the following reason. The nth subquotients in the chromatic filtration consists of v_n-periodic elements. As illustrated in 2.4.2, these elements fall into periodic families. The chromatic filtration is thus like a spectrum in the astronomical sense in that it resolves the stable homotopy groups of a finite complex into periodic families of various periods. Comparing these to the colors of the rainbow led us to the word 'chromatic.'

The construction outlined above differs slightly from that used in 2.4.2. Suppose for example that we apply chromatic analysis to the map

$$\alpha_1 : S^{m+2p-3} \to S^m$$

for an odd prime p. This element in π_{q-1}^S has order p so the map extends to the mod p Moore space $V(0)_{m+q-1}$, which has the Adams self-map α of

2.4.1(ii). We find that all iterates of α when composed with α_1 are essential, so α_1 is v_1-periodic. The composite

$$S^{m+qi+q-1} \longrightarrow V(0)_{m+qi+q-1} \xrightarrow{\alpha^i} V(0)_{m+q-1} \xrightarrow{\tilde{\alpha}_1} S^m$$

is the map α_{i+1} of 2.4.1(ii). In the chromatic analysis no use is made of the map $j: V(0)_m \to S^{m+1}$.

More generally, suppose $g: S^k \to Y$ is v_n-periodic and that it extends to $g_n: W(n) \to Y$. There is no guarantee that the composite

$$S^K \xrightarrow{e} \Sigma^{d_n i} W(n) \xrightarrow{f_n^i} W(n) \xrightarrow{g_n} Y$$

(where e is the inclusion of the bottom cell in $W(n)$) is essential, even though $g_n f_n^i$ is essential by assumption. (This is the case in each of the examples of 2.4.2.) If this composite is null homotopic then $g_n f_n^i$ extends to the cofibre of e. Again, this extension may or may not be essential on the bottom cell of C_e. However, $g_n f_n^i$ must be nontrivial on one of the 2^n cells of $W(n)$ since it is an essential map. (To see this, one can make a construction similar to that shown in (2.4.4) and (2.4.5).) Thus for each i we get some nontrivial element in $\pi_*^S(Y)$.

Definition 2.5.3 *Given a v_n-periodic element $y \in \pi_*^S(Y)$, the elements described above for various $i > 0$ constitute the v_n-**periodic** family associated with y.*

One can ask if the chromatic analysis of a given element terminates after a finite number of steps. For a reformulation of this question, see the chromatic convergence theorem, 7.5.7

Chapter 3

MU-theory and formal group laws

In this section we will discuss the homology theory MU_* used in the nilpotence theorem. $MU_*(X)$ is defined in terms of maps of manifolds into X as will be explained presently. Unfortunately the geometry in this definition does not appear to be relevant to the applications we have in mind. We will be more concerned with some algebraic properties of the functor which are intimately related to the theory of formal group laws.

3.1 Complex bordism

Definition 3.1.1 *Let M_1 and M_2 be smooth closed n-dimensional manifolds, and let $f_i: M_i \to X$ be continuous maps for $i = 1, 2$. These maps are* **bordant** *if there is a map $f: W \to X$, where W is a smooth manifold whose boundary is the disjoint union of M_1 and M_2, such that the restriction of f to M_i is f_i. f is a* **bordism** *between f_1 and f_2.*

Bordism is an equivalence relation and the set of bordism classes forms a group under disjoint union, called the nth **bordism group** of X.

A manifold is **stably complex** if it admits a complex linear structure in its stable normal bundle, i.e., the normal bundle obtained by embedding in a large dimensional Euclidean space. (The term stably *almost* complex is often used in the literature.) A complex analytic manifold (e.g. a complex algebraic variety) is stably complex, but the notion of stably complex is far weaker than that of complex analytic.

Definition 3.1.2 $MU_n(X)$, *the nth* **complex bordism group** *of X, is the bordism group obtained by requiring that all manifolds in sight be stably* *complex.*

The fact that these groups are accessible is due to some remarkable work of Thom in the 1950's [Tho54]. More details can be found below in B.2. A general reference for cobordism theory is Stong's book [Sto68]. More information can be found below in B.2.

The groups $MU_*(X)$ satisfy all but one of the axioms used by Eilenberg-Steenrod to characterize ordinary homology; see A.3. They fail to satisfy the dimension axiom, which describes the homology of a point. If X is a single point, then the map from the manifold to X is unique, and $MU_*(\text{pt.})$ is the group of bordism classes of stably complex manifolds, which we will denote simply by MU_*. It is a graded ring under Cartesian product and its structure was determined independently by Milnor [Mil60] and Novikov [Nov60] and [Nov62].

Theorem 3.1.3 *The complex bordism ring, MU_* is isomorphic to*

$$\mathbf{Z}[x_1, x_2, \ldots]$$

where $\dim x_i = 2i$.

It is possible to describe the generators x_i as complex manifolds, but this is more trouble than it is worth. The complex projective spaces $\mathbf{C}P^i$ serve as polynomial generators of $\mathbf{Q} \otimes MU_*$.

Note that $MU_*(X)$ is an MU_*-module as follows. Given $x \in MU_*(X)$ represented by $f \colon M \to X$ and $\lambda \in MU_*$ represented by a manifold N, λx is represented by the composite map

$$M \times N \longrightarrow M \xrightarrow{f} X.$$

3.2 Formal group laws

Definition 3.2.1 *A formal group law over a commutative ring with unit R is a power series $F(x, y)$ over R that satisfies the following three conditions.*
  (i) $F(x, 0) = F(0, x) = x$ (identity),
  (ii) $F(x, y) = F(y, x)$ (commutativity) and
  (iii) $F(F(x, y), z) = F(x, F(y, z))$ (associativity).
(The existence of an inverse is automatic. It is the power series $i(x)$ determined by the equation $F(x, i(x)) = 0$.)

Example 3.2.2 *(i)* $F(x, y) = x + y$. *This is called the* **additive** *formal group law.*

(ii) $F(x, y) = x + y + xy = (1 + x)(1 + y) - 1$. *This is called the* **multiplicative** *formal group law.*

(iii)

$$F(x, y) = \frac{x\sqrt{R(y)} + y\sqrt{R(x)}}{1 - \varepsilon x^2 y^2}$$

where

$$R(x) = 1 - 2\delta x^2 + \varepsilon x^4.$$

This is the formal group law associated with the elliptic curve

$$y^2 = R(x),$$

a Jacobi quartic, so we call it the **elliptic** *formal group law. It is defined over* $\mathbf{Z}[1/2][\delta, \varepsilon]$. *This curve is nonsingular mod p (for p odd) if the discriminant* $\Delta = \varepsilon(\delta^2 - \varepsilon)^2$ *is invertible. This example figures prominently in elliptic cohomology theory; see [LRS] for more details.*

The theory of formal group laws from the power series point of view is treated comprehensively in [Haz78]. A short account containing all that is relevant for the current discussion can be found in [Rav86, Appendix 2].

The following result is due to Lazard [Laz55a].

Theorem 3.2.3 (Lazard's theorem) *(i) There is a universal formal group law defined over a ring L of the form*

$$G(x, y) = \sum_{i,j} a_{i,j} x^i y^j \quad \text{with } a_{i,j} \in L$$

such that for any formal group law F over R there is a unique ring homomorphism θ *from L to R such that*

$$F(x, y) = \sum_{i,j} \theta(a_{i,j}) x^i y^j.$$

(ii) L is a polynomial algebra $\mathbf{Z}[x_1, x_2, \ldots]$. *If we put a grading on L such that* $a_{i,j}$ *has degree* $2(1 - i - j)$ *then* x_i *has degree* $-2i$.

The grading above is chosen so that if x and y have degree 2, then $G(x, y)$ is a homogeneous expression of degree 2. Note that L is isomorphic to MU_* except that the grading is reversed. There is an important connection between the two.

Associated with the homology theory MU_* there is a cohomology theory MU^*. This is a contravariant functor bearing the same relation to MU_*

that ordinary cohomology bears to ordinary homology. When X is an m-dimensional manifold, $MU^*(X)$ has a geometric description; an element in $MU^k(X)$ is represented by a map to X from an $(m - k)$-dimensional manifold with certain properties. The conventions in force in algebraic topology require that $MU^*(\text{pt.})$ (which we will denote by MU^*) be the same as $MU_*(\text{pt.})$ but with the grading reversed. Thus MU^* is isomorphic to the Lazard ring L.

This isomorphism is natural in the following sense. $MU^*(X)$, like $H^*(X)$, comes equipped with cup products, making it a graded algebra over MU^*. Of particular interest is the case when X is the infinite-dimensional complex projective space $\mathbf{C}P^\infty$. We have

$$MU^*(\mathbf{C}P^\infty) \cong MU^*[[x]]$$

where dim $x = 2$, and

$$MU^*(\mathbf{C}P^\infty \times \mathbf{C}P^\infty) \cong MU^*[[x \otimes 1, 1 \otimes x]].$$

The space $\mathbf{C}P^\infty$ is an abelian topological group, so there is a map

$$\mathbf{C}P^\infty \times \mathbf{C}P^\infty \xrightarrow{\ f\ } \mathbf{C}P^\infty$$

with certain properties. ($\mathbf{C}P^\infty$ is also the classifying space for complex line bundles and the map in question corresponds to the tensor product.) Since MU^* is contravariant we get a map

$$MU^*(\mathbf{C}P^\infty \times \mathbf{C}P^\infty) \xleftarrow{\ f^*\ } MU^*(\mathbf{C}P^\infty)$$

which is determined by its behavior on the generator $x \in MU^2(\mathbf{C}P^\infty)$. The power series

$$f^*(x) = F(x \otimes 1, 1 \otimes x)$$

can easily be shown to be a formal group law. Hence by Lazard's theorem (3.2.3) it corresponds to a ring homomorphism $\theta \colon L \to MU^*$. The following was proved by Quillen [Qui69] in 1969.

Theorem 3.2.4 (Quillen's theorem) *The homomorphism*

$$\theta \colon L \to MU^*$$

above is an isomorphism. In other words, the formal group law associated with complex cobordism is the universal one.

Given this isomorphism (and ignoring the reversal of the grading), we can regard $MU_*(X)$ as an L-module.

3.3 The category CΓ

Now we define a group Γ which acts in an interesting way on L.

Definition 3.3.1 *Let Γ be the group of power series over $\mathbf{Z}$ having the form*

$$\gamma = x + b_1 x^2 + b_2 x^3 + \cdots$$

where the group operation is functional composition. Γ acts on the Lazard ring L of 1.5 as follows. Let $G(x,y)$ be the universal formal group law as above and let $\gamma \in \Gamma$. Then $\gamma^{-1}(G(\gamma(x),\gamma(y)))$ is another formal group law over L, and therefore is induced by a homomorphism from L to itself. Since γ is invertible, this homomorphism is an automorphism, giving the desired action of Γ on L.

For reasons too difficult to explain here, Γ also acts naturally on $MU_*(X)$ compatibly with the action on $MU_*(\text{pt.})$ defined above. (For more information about this, see B.3 and B.4.) That is, given $x \in MU_*(X)$, $\gamma \in \Gamma$ and $\lambda \in L$, we have

$$\gamma(\lambda x) = \gamma(\lambda)\gamma(x)$$

and the action of Γ commutes with homomorphisms induced by continuous maps.

For algebraic topologists we can offer some explanation for this action of Γ. It is analogous to the action of the Steenrod algebra in ordinary cohomology. More precisely, it is analogous to the action of the group of multiplicative cohomology operations, such as (in the mod 2 case) the total Steenrod square, $\sum_{i \geq 0} \text{Sq}^i$. Such an operation is determined by its effect on the generator of $H^1(\mathbf{R}P^\infty; \mathbf{Z}/(2))$. Thus the group of multiplicative mod 2 cohomology operations embeds in $\Gamma_{\mathbf{Z}/(2))}$, the group of power series over $\mathbf{Z}/(2)$ analogous to Γ over the integers.

Definition 3.3.2 *Let CΓ denote the category of finitely presented graded L-modules equipped with an action of Γ compatible with its action on L as above, and let **FH** denote the category of finite CW-complexes and homotopy classes of maps between them.*

Thus we can regard $\overline{MU}_*$ as a functor from **FH** to CΓ. The latter category is much more accessible. We will see that it has some structural features which reflect those of **FH** very well. The nilpotence, periodicity and chromatic convergence theorems are examples of this.

In order to study CΓ further we need some more facts about formal group laws. Here are some power series associated with them.

Definition 3.3.3 *For each integer n the n-series $[n](x)$ is given by*

$$\begin{aligned}
[1](x) &= x, \\
[n](x) &= F(x, [n-1](x)) \qquad \text{for } n > 1 \text{ and} \\
[-n](x) &= i([n](x)).
\end{aligned}$$

These satisfy

$$\begin{aligned}
[n](x) &\equiv nx \quad \bmod (x^2), \\
[m+n](x) &= F([m](x), [n](x)) \qquad \text{and} \\
[mn](x) &= [m]([n](x)).
\end{aligned}$$

For the additive formal group law(3.2.2), we have $[n](x) = nx$, and for the multiplicative formal group law, $[n](x) = (1 + x)^n - 1$.

Of particular interest is the p-series. In characteristic p it always has leading term ax^q where $q = p^h$ for some integer h. This leads to the following.

Definition 3.3.4 *Let $F(x, y)$ be a formal group law over a ring in which the prime p is not a unit. If the mod p reduction of $[p](x)$ has the form*

$$[p](x) = ax^{p^h} + \text{higher terms}$$

*with a invertible, then we say that F has **height h** at p. If $[p](x) \equiv 0 \bmod p$ then the height is infinity.*

For the additive formal group law we have $[p](x) = 0$ so the height is ∞. The multiplicative formal group law has height 1 since $[p](x) = x^p$. The mod p reduction (for p odd) of the elliptic formal group law of 3.2.2(iii) has height one or two depending on the values of δ and ε. For example if $\delta = 0$ and $\varepsilon = 1$ then the height is one for $p \equiv 1 \bmod 4$ and two for $p \equiv 3 \bmod 4$. (See [Rav86, pages 373–374])

The following classification theorem is due to Lazard [Laz55b].

Theorem 3.3.5 (Classification of formal group laws) *Two formal group laws over the algebraic closure of $\mathbf{F}_p$ are isomorphic if and only if they have the same height.*

Let $v_n \in L$ denote the coefficient of x^{p^n} in the p-series for the universal formal group law; the prime p is omitted from the notation. This v_n is closely related to the v_n in the Morava K-theories (1.5.2); the precise relation is explained in B.7. It can be shown that v_n is an indecomposable element in L, i.e., it could serve as a polynomial generator in dimension $2p^n - 2$. Let $I_{p,n} \subset L$ denote the prime ideal $(p, v_1, \ldots v_{n-1})$.

The following result is due to Morava [Mor85] and Landweber [Lan73a].

Theorem 3.3.6 (Invariant prime ideal theorem) *The only prime ideals in L which are invariant under the action of Γ are the $I_{p,n}$ defined above, where p is a prime integer and n is a nonnegative integer, possibly ∞. ($I_{p,\infty}$ is by definition the ideal $(p, v_1, v_2, \ldots)$ and $I_{p,0}$ is the zero ideal.)*

Moreover in $L/I_{p,n}$ for $n > 0$ the subgroup fixed by Γ is $\mathbf{Z}/(p)[v_n]$. In L itself the invariant subgroup is $\mathbf{Z}$.

This shows that the action of Γ on L is very rigid. L has a bewildering collection of prime ideals, but the only ones we ever have to consider are the ones listed in the theorem. This places severe restriction on the structure of modules in CΓ.

Recall that a finitely presented module M over a Noetherian ring R has a finite filtration

$$F_1 M \subset F_2 M \subset \cdots F_k M = M$$

in which each subquotient $F_i M / F_{i-1} M$ is isomorphic to R/I_i for some prime ideal $I_i \subset R$. Now L is not Noetherian, but it is coherent, which means that finitely presented modules over it admit similar filtrations. For a module in CΓ, the filtration can be chosen so that the submodules, and therefore the prime ideals, are all invariant under Γ. The following result is due to Landweber [Lan73b].

Theorem 3.3.7 (Landweber filtration theorem) *Every module M in CΓ admits a finite filtration by submodules in CΓ as above in which each subquotient is isomorphic to a suspension (recall that the modules are graded) of $L/I_{p,n}$ for some prime p and some finite n.*

These results suggest that, once we have localized at a prime p, the only polynomial generators of MU_* which really matter are the $v_n = x_{p^n - 1}$. In fact the other generators act freely on any module in CΓ and hence provide no information. We might as well tensor them away and replace the theory of L-modules with Γ-action by a corresponding theory of modules over the ring

$$V_p = \mathbf{Z}_{(p)}[v_1, v_2, \cdots]. \tag{3.3.8}$$

This has been done and the ring V_p is commonly known as BP_*, the coefficient ring for Brown-Peterson theory. There are good reasons for doing this from the topological standpoint, from the formal group law theoretic standpoint, and for the purpose of making explicit calculations useful in homotopy theory. Indeed all of the current literature on this subject is written in terms of BP-theory rather than MU-theory.

However it is *not* necessary to use this language in order to describe the subject conceptually as we are doing here. Hence we will confine our treatment of BP to the Appendix (B.5). There is one technical problem with BP-theory which makes it awkward to discuss in general terms. There is no BP-theoretic analogue of the group Γ. It has to be replaced instead by a certain groupoid, and certain Hopf algebras associated with MU-theory have to be replaced by Hopf algebroids (see B.3).

The following are easy consequences of the Landweber filtration theorem.

Corollary 3.3.9 *Suppose M is a p-local module in $\mathbf{C\Gamma}$ and $x \in M$.*

(i) If x is annihlated by some power of v_n, then it is annihilated by some power of v_{n-1}, so if $v_n^{-1}M = 0$, i.e., if each element in M is annihilated by some power of v_n, then $v_{n-1}^{-1}M = 0$.

(ii) If x is nontrivial, then there is an n so that $v_n^k x \neq 0$ for all k, so if M is nontrivial, then so is $v_n^{-1}M$ for all sufficiently large n.

(iii) If $v_{n-1}^{-1}M = 0$, then there is a positive integer k such that multiplication by v_n^k in M commutes with the action of Γ.

(iv) Conversely, if $v_{n-1}^{-1}M$ is nontrivial, then there is no positive integer k such that multiplication by v_n^k in M commutes with the action of Γ on x.

The first two statements should be compared to the last two statements in 1.5.2. In fact the functor $v_n^{-1}\overline{MU}_*(X)_{(p)}$ is a homology theory (see B.6.2) which vanishes on a finite p-local CW-complex X if and only if $\overline{K(n)}_*(X)$ does. One could replace $K(n)_*$ by $v_n^{-1}MU_{(p)*}$ in the statement of the periodicity theorem. The third statement is an algebraic analogue of the periodicity theorem.

We can mimic the definition of type n finite spectra (1.5.3) and v_n-maps (1.5.4) in $\mathbf{C\Gamma}$.

Definition 3.3.10 *A p-local module M in $\mathbf{C\Gamma}$ has type n if n is the smallest integer with $v_n^{-1}M$ nontrivial. A homomorphism $f : \Sigma^d M \to M$ in $\mathbf{C\Gamma}$ is a v_n-map if it induces an isomorphism in $v_n^{-1}M$ and the trivial homomorphism in $v_m^{-1}M$ for $m \neq n$.*

The another consequence of the Landweber filtration theorem is the following.

Corollary 3.3.11 *If M in $\mathbf{C\Gamma}$ is a p-local module with $v_{n-1}^{-1}M$ nontrivial, then M does not admit a v_n-map.*

Sketch of proof of 3.3.9. (i) The statement about x is proved by Johnson-Yosimura in [JY80]. The statement about M can be proved independently as follows. The condition implies that each subquotient in the Landweber filtration is a suspension of $L/I_{p,m}$ for some $m > n$. It follows that each element is annihilated by some power of v_{n-1} as claimed.

(ii) We can choose n so that each Landweber subquotient of M is a suspension of $L/I_{p,m}$ for some $m \leq n$. Then no element of M is annihilated by any power of v_n.

(iii) If $v_{n-1}^{-1}M = 0$, then each Landweber subquotient is a suspension of $L/I_{p,m}$ for $m \geq n$. It follows that if the length of the filtration is j, then M is annihilated by $I_{p,n}^j$. For any $\gamma \in \Gamma$ we have

$$\gamma(v_n) = v_n + e \quad \text{with} \quad e \in I_{p,n}.$$

It follows easily that

$$
\begin{aligned}
\gamma(v_n^{p^{j-1}}) &= (v_n + e)^{p^{j-1}} \\
&= v_n^{p^{j-1}} + e' \quad \text{with} \quad e' \in I_{p,n}^j.
\end{aligned}
$$

This means that multiplication by $v_n^{p^{j-1}}$ is Γ-equivariant in $L/I_{p,n}^j$ and hence in M.

(iv) Suppose such an integer k exists. Then multiplication by v_n^k is Γ-equivariant on each Landweber subquotient. However by 3.3.6 this is not the case on $L/I_{p,m}$ for $m < n$. It follows that $v_{n-1}^{-1}M = 0$, which is a contradiction. ∎

Proof of 3.3.11. Suppose M has type m for $m < n$. This means that each Landweber subquotient of M is a suspension of $L/I_{p,k}$ for some $k \geq m$. Hence we see that $v_m^{-1}M$, $v_n^{-1}M$ and hence $v_m^{-1}v_n^{-1}M$ are all nontrivial. On the other hand, if f is a v_n-map, then $v_m^{-1}v_n^{-1}f$ must be both trivial and an isomorphism, which is a contradiction. ∎

3.4 Thick subcategories

Now we need to consider certain full subcategories of $\mathbf{C\Gamma}$ and $\mathbf{FH}$.

Definition 3.4.1 *A full subcategory* $\mathbf{C}$ *of* $\mathbf{C\Gamma}$ *is* **thick** *if it satisfies the following axiom:*
If

$$0 \longrightarrow M' \longrightarrow M \longrightarrow M'' \longrightarrow 0$$

is a short exact sequence in $\mathbf{C\Gamma}$, *then* M *is in* $\mathbf{C}$ *if and only if* M' *and* M'' *are.* *(In other words* $\mathbf{C}$ *is closed under subobjects, quotient objects, and extensions.)*

A full subcategory $\mathbf{F}$ *of* $\mathbf{FH}$ *is thick if it satisfies the following two axioms:*

(i) If

$$X \xrightarrow{f} Y \longrightarrow C_f$$

is a cofibre sequence in which two of the three spaces are in **F**, *then so is the third.*

(ii) If $X \vee Y$ *is in* **F** *then so are* X *and* Y.

Thick subcategories were called generic subcategories by Hopkins in [Hop87].

Using the Landweber filtration theorem, one can classify the thick subcategories of $C\Gamma_{(p)}$.

Theorem 3.4.2 *Let* **C** *be a thick subcategory of* $C\Gamma_{(p)}$ *(the category of all p-local modules* $C\Gamma$*). Then* **C** *is either all of* $C\Gamma_{(p)}$, *the trivial subcategory (in which the only object is the trivial module), or consists of all p-local modules M in* $C\Gamma$ *with* $v_{n-1}^{-1} M = 0$. *We denote the latter category by* $C_{p,n}$.

We will sketch the proof of this result below.

There is an analogous result about thick subcategories of $FH_{(p)}$, which is a very useful consequence of the nilpotence theorem.

Theorem 3.4.3 (Thick subcategory theorem) *Let* **F** *be a thick subcategory of* $FH_{(p)}$, *the category of p-local finite CW-complexes. Then* **F** *is either all of* $FH_{(p)}$, *the trivial subcategory (in which the only object is a point) or consists of all p-local finite CW-complexes X with* $v_{n-1}^{-1} \overline{MU}_*(X) = 0$. *We denote the latter category by* $F_{p,n}$.

Thus we have two nested sequences of thick subcategories,

$$FH_{(p)} = F_{p,0} \supset F_{p,1} \supset F_{p,2} \cdots \{\text{pt.}\} \tag{3.4.4}$$

and

$$C\Gamma_{(p)} = C_{p,0} \supset C_{p,1} \supset C_{p,2} \cdots \{0\}. \tag{3.4.5}$$

The functor $MU_*(\cdot)$ sends one to the other. Until 1983 it was not even known that the $F_{p,n}$ were nontrivial for all but a few small values of n. Mitchell [Mit85] first showed that all of the inclusions of the $F_{p,n}$ are proper. Now it is a corollary of the periodicity theorem.

In Chapter 4 we will describe another algebraic paradigm analogous to 3.4.3 discovered in the early 70's by Jack Morava. It points to some

interesting connections with number theory and was the original inspiration
behind this circle of ideas.

In Chapter 5 we will derive the thick subcategory theorem from another
form of the nilpotence theorem. This is easy since it uses nothing more than
elementary tools from homotopy theory.

In Chapter 6 we will sketch the proof of the periodicity theorem. It is
not difficult to show that the collection of complexes admitting periodic self
maps for given p and n forms a thick subcategory. Given the thick subcategory theorem, it suffices to find just one nontrivial example of a complex of
type n with a periodic self-map. This involves some hard homotopy theory.
There are two major ingredients in the construction. One is the Adams
spectral sequence, a computational tool that one would expect to see used
in such a situation. The other is a novel application of the modular representation theory of the symmetric group described in as yet unpublished
work of Jeff Smith.

Now we will discuss the proof of 3.4.2.

Given two objects M and N in $\mathbf{C\Gamma}$, we can define a Γ-actions on
$M \otimes_{MU_*} N$ (denoted hereafter simply by $M \otimes N$) by

$$\gamma(m \otimes n) = \gamma(m) \otimes \gamma(n)$$

and on $\mathrm{Hom}_{MU_*}(M, N)$ (denoted hereafter by $\mathrm{Hom}(M, N)$) by

$$\gamma(f)(m) = \gamma(f(\gamma^{-1}(m))).$$

Note that the homomorphisms in $\mathrm{Hom}(M, N)$ are *not* required to be Γ-
equivariant.

Proposition 3.4.6 *If* $\mathbf{C} \subset \mathbf{C\Gamma}$ *is thick (3.4.1) and* M *is in* $\mathbf{C}$, *then so are*
$N \otimes M$ *and* $\mathrm{Hom}(N, M)$ *for any* N *in* $\mathbf{C\Gamma}$.

Sketch of proof. Recall that each object N in $\mathbf{C\Gamma}$ is finitely presented as
an L-module. A consequence of the Landweber filtration theorem (Theo-
rem 3.3.7) is that each N has a finite free resolution in $\mathbf{C\Gamma}$, i.e., an exact
sequence of the form

$$0 \longleftarrow N \longleftarrow F_0 \longleftarrow F_1 \longleftarrow \cdots F_n \longleftarrow 0$$

where each F_i is free and finitely generated over L.

Now if M is in $\mathbf{C}$, then so are $F_i \otimes M$ and $\mathrm{Hom}(F_i, M)$ for each i. Using
this, one can show by induction on n that $N \otimes M$ and $\mathrm{Hom}(N, M)$ are in
$\mathbf{C}$ as claimed. ∎

Now suppose $\mathbf{C} \subset \mathbf{C\Gamma}_{(p)}$ is thick. Choose the smallest n so that $\mathbf{C}_{p,n} \supset$
$\mathbf{C}$. We want to show that $\mathbf{C} \supset \mathbf{C}_{p,n}$. Let M be in $\mathbf{C}$ but not in $\mathbf{C}_{p,n+1}$.

Then its endomorphism ring

$$\text{End}(M) = \text{Hom}(M, M)$$

is in $\mathbf{C}$ by 3.4.6. There is a natural map

$$MU_{(p)*} \xrightarrow{\ s\ } \text{End}(M)$$

Its kernel is the *annihilator ideal of M*, which we denote by $\text{Ann}(M)$. It is Γ-invariant. (For more discussion of annihilator ideals, see Landweber's paper [Lan79].)

It was shown in the proof of 3.3.9(iii) above that $\text{Ann}(M)$ must contain some power of $I_{p,n}$. On the other hand, it cannot contain any power of v_m for $m \geq n$, so it is contained in $I_{p,n}$.

Similar statements can be made about $\text{Ann}(N)$ for any N in $\mathbf{C}_{p,n}$. (It is actually enough to consider only the case $N = MU_*/I_{p,n}$.) Hence we can choose k so that

$$\text{Ann}(N) \supset \text{Ann}(M)^k. \tag{3.4.7}$$

Now consider the exact sequence

$$0 \longrightarrow \text{Ann}(M)^k \longrightarrow MU_{(p)*} \xrightarrow{\ s^{\otimes k}\ } \text{End}(M)^{\otimes k}$$

It follows from (3.4.7) that tensoring with N gives a monomorphism

$$0 \longrightarrow N \longrightarrow N \otimes \text{End}(M)^{\otimes k}.$$

The target is in $\mathbf{C}$ by 3.4.6, so N is in $\mathbf{C}$, and $\mathbf{C} = \mathbf{C}_{p,n}$ as desired.

Chapter 4

Morava's orbit picture and Morava stabilizer groups

In this section we will describe some ideas conceived by Jack Morava in the early 70's. It is a pleasure to acknowledge once again his inspiring role in this area.

4.1 The action of Γ on L

The action of the group Γ on the Lazard ring L (3.3.1) is central to this theory and the picture we will describe here sheds considerable light on it. Let $H_{\mathbf{Z}}L$ denote the set of ring homomorphisms $L \to \mathbf{Z}$. By 3.2.3 this is the set of formal group laws over the integers. Since L is a polynomial ring, a homomorphism $\theta \in H_{\mathbf{Z}}L$ is determined by its values on the polynomial generators $x_i \in L$. Hence $H_{\mathbf{Z}}L$ can be regarded as an infinite dimensional affine space over $\mathbf{Z}$. The action of Γ on L induces one on $H_{\mathbf{Z}}L$. The following facts about it are straightforward.

Proposition 4.1.1 *Let $H_{\mathbf{Z}}L$ and the action of the group Γ on it be as above. Then*

(i) Points in $H_{\mathbf{Z}}L$ correspond to formal group laws over $\mathbf{Z}$.

(ii) Two points are in the same Γ-orbit if and only if the two corresponding formal group laws are isomorphic over $\mathbf{Z}$.

(iii) The subgroup of Γ fixing point $\theta \in H_{\mathbf{Z}}L$ is the strict automorphism group of the corresponding formal group law.

(iv) The strict automorphism groups of isomorphic formal group laws are conjugate in Γ.

We have not yet said what a strict automorphism of a formal group law F is.

An automorphism is a power series $f(x)$ satisfying

$$f(F(x,y)) = F(f(x), f(y))$$

and $f(x)$ is *strict* if it has the form

$$f(x) = x + \text{ higher terms.}$$

The classification of formal group laws over the integers is quite complicated, but we have a nice classification theorem (3.3.5) over k, the algebraic closure of $\mathbf{F}_p$. Hence we want to replace $\mathbf{Z}$ by k in the discussion above. Let $H_k L$ denote the set of ring homomorphisms $L \to k$; it can be regarded as an infinite dimensional vector space over k. Let Γ_k denote the corresponding group of power series. Then it follows that there is one Γ_k-orbit for each height n. Since $\theta(v_i) \in k$ is the coefficient of x_{p^i} in the power series $[p](x)$, the following is a consequence of the relevant definitions.

Proposition 4.1.2 *The formal group law over k corresponding to $\theta \in H_k L$ has height n if and only if $\theta(v_i) = 0$ for $i < n$ and $\theta(v_n) \neq 0$. Moreover, each $v_n \in L$ is indecomposable, i.e., it is a unit (in $\mathbf{Z}_{(p)}$) multiple of*

$$x_{p^n-1} + \text{ decomposables.}$$

Let $Y_n \subset H_k L$ denote the height n orbit. It is the subset defined by the equations $v_i = 0$ for $i < n$ and $v_n \neq 0$ for finite n, and for $n = \infty$ it is defined by $v_i = 0$ for all $n < \infty$. Let

$$X_n = \bigcup_{n \leq i} Y_i$$

so we have a nested sequence of subsets

$$H_k L = X_1 \supset X_2 \supset X_3 \cdots X_\infty \tag{4.1.3}$$

which is analogous to (3.4.4) and (3.4.5).

4.2 Morava stabilizer groups

Now we want to describe the strict automorphism group S_n (called the nth **Morava stabilizer group**) of a height n formal group law over k. It is contained in the multiplicative group over a certain division algebra D_n over the p-adic numbers $\mathbf{Q}_p$. To describe it we need to define several other algebraic objects.

Recall that $\mathbf{F}_{p^n}$, the field with p^n elements, is obtained from $\mathbf{F}_p$ by adjoining a primitive $(p^n - 1)^{\text{st}}$ root of unity $\overline{\zeta}$, which is the root of some irreducible polynomial of degree n. The Galois group of this extension is cyclic of order n generated by the Frobenius automorphism which sends an element x to x^p.

There is a corresponding degree n extension $W(\mathbf{F}_{p^n})$ of the p-adic integers $\mathbf{Z}_p$, obtained by adjoining a primitive $(p^n - 1)^{\text{st}}$ root of unity ζ (whose mod p reduction is $\overline{\zeta}$), which is also the root of some irreducible polynomial of degree n. The Frobenius automorphism has a lifting σ fixing $\mathbf{Z}_p$ with $\sigma(\zeta) = \zeta^p$ and

$$\sigma(x) \equiv x^p \bmod p$$

for any $x \in W(\mathbf{F}_{p^n})$.

We denote the fraction field of $W(\mathbf{F}_{p^n})$ by K_n; it is the unique unramified extension of $\mathbf{Q}_p$ of degree n. Let $K_n\langle S \rangle$ denote the ring obtained by adjoining a *noncommuting* power series variable S subject to the rule

$$Sx = \sigma(x)S$$

for $x \in K_n$. Thus S commutes with everything in $\mathbf{Q}_p$ and S^n commutes with all of K_n. The division algebra D_n is defined by

$$D_n = K_n\langle S \rangle / (S^n - p). \tag{4.2.1}$$

It is an algebra over $\mathbf{Q}_p$ of rank n^2 with center $\mathbf{Q}_p$. It is known to contain each degree n field extension of $\mathbf{Q}_p$ as a subfield. (This statement is 6.2.12 of [Rav86], where appropriate references are given.)

It also contains a maximal order

$$E_n = W(\mathbf{F}_{p^n})\langle S \rangle / (S^n - p). \tag{4.2.2}$$

E_n is a complete local ring with maximal ideal (S) and residue field $\mathbf{F}_{p^n}$. Each element in $a \in E_n$ can be written uniquely as

$$a = \sum_{0 \le i \le n-1} a_i S^i \tag{4.2.3}$$

with $a_i \in W(\mathbf{F}_{p^n})$, and also as

$$a = \sum_{i \geq 0} e_i S^i \tag{4.2.4}$$

where each $e_i \in W(\mathbf{F}_{p^n})$ satisfies the equation

$$e_i^{p^n} - e_i = 0,$$

i.e., e_i is either zero or a root of unity. The groups of units $E_n^\times \subset E_n$ is the set of elements with $e_0 \neq 0$, or equivalently with a_0 a unit in $W(\mathbf{F}_{p^n})$.

Proposition 4.2.5 *The full automorphism group of a formal group law over k of height n is isomorphic to $E_n^\times$ above, and the strict automorphism group S_n is isomorphic to the subgroup of $E_n^\times$ with $e_0 = 1$.*

If we regard each coefficient e_i as a continuous $\mathbf{F}_{p^n}$-valued function on S_n, then it can be shown that the ring of all such functions is

$$S(n) = \mathbf{F}_{p^n}[e_i : i \geq 1]/(e_i^{p^n} - e_i). \tag{4.2.6}$$

This is a Hopf algebra over $\mathbf{F}_{p^n}$ with coproduct induced by the group structure of S_n. It should be compared to the Hopf algebra $\Sigma(n)$ of (B.7.5). This is a factor of $K(n)_*(K(n))$, the Morava K-theory analog of the dual Steenrod algebra. Its multiplicative structure is given by

$$\Sigma(n) = K(n)_*[t_1, t_2, \cdots]/(t_i^{p^n} - v_n^{p^i - 1} t_i).$$

Hence we have
$$S(n) = \Sigma(n) \otimes_{K(n)_*} \mathbf{F}_{p^n}$$

under the isomorphism sending t_i to e_i and v_n to 1.

Now we will describe the action of S_n on a particular height n formal group law F_n. To define F_n, let F be the formal group law over $\mathbf{Z}_{(p)}$ with logarithm

$$\log_F(x) = \sum_{i \geq 0} \frac{x^{p^{in}}}{p^i}. \tag{4.2.7}$$

F_n is obtained by reducing F mod p and tensoring with $\mathbf{F}_{p^n}$.

Now an automorphism e of F_n is a power series $e(x)$ over $\mathbf{F}_{p^n}$ satisfying

$$e(F_n(x, y)) = F_n(e(x), e(y)).$$

For

$$e = \sum_{i \geq 0} e_i S^i \in S_n$$

(with $e_0 = 1$) we define $e(x)$ by

$$e(x) = \sum_{i \geq 0} {}^{F_n} e_i x^{p^i}. \tag{4.2.8}$$

More details can be found in [Rav86, Appendix 2].

4.3 Cohomological properties of S_n

We will see below in Sections 7.5, 8.3 and 8.4 that the cohomology of the group S_n figures prominently in the stable homotopy groups of finite complexes. For future reference we will record some facts about this cohomology here. Proofs and more precise statements can be found in [Rav86, Chapter 6].

First we will say something about why this cohomology is relevant. At the beginning of 3.3 we remarked that the group Γ is essentially the group of multiplicative cohomology operations in MU-theory. The same can be said of S_n in Morava K-theory. More precisely, consider the functor

$$FK(n)_*(X) = K(n)_*(X) \otimes_{K(n)_*} \mathbf{F}_{p^n}, \tag{4.3.1}$$

where $\mathbf{F}_{p^n}$ (the field with p^n elements) is made into a $K(n)_*$-module by sending v_n to 1. Hence we must ignore the grading in order to define this tensor product. $FK(n)_*$ takes values in the category of $\mathbf{Z}/(2)$-graded vector spaces over $\mathbf{F}_{p^n}$. It can be shown that *the group of multiplicative operations in this theory is precisely* S_n. The field $\mathbf{F}_{p^n}$ is essential here; if we were to replace it by the prime field, we would not have the same result. Moreover, replacing it by a larger field would not enlarge the group of multiplicative operations.

In ordinary cohomology and in MU-theory, one cannot recover the action of the full algebra of cohomology operations from that of the multiplicative operations. However, one can do this in Morava K-theory, after making suitable allowances for the Bockstein operation. Classically one uses ordinary cohomology operations to compute homotopy groups via the Adams spectral sequence; see A.6 for a brief introduction. This requires the computations of various Ext groups over the algebra of cohomology operations. Analogous computations in Morava K-theory amount to finding the cohomology of S_n with various coefficients.

S_n is a profinite group, which means that its topology must be taken into account in order to do sensible cohomological computations. It can also be described as a "p-adic Lie group;" see [Laz65]. These technicalities can be avoided by formulating its cohomology in Hopf-algebra theoretic terms. For now we will simply write $H^*(S_n)$ to denote the continuous mod p cohomology of S_n and refer the interested reader to [Rav86, Chapter 6] for a precise definition. $H^*(S_n)$ is computed there explicitly at all primes for $n = 1$ and 2 at all primes, and for $n = 3$ when $p \geq 5$.

Theorem 4.3.2 *(a) $H^*(S_n)$ is finitely generated as an algebra.*

(b) If n is not divisible by $p - 1$, then $H^i(S_n) = 0$ for $i > n^2$, and for $0 \leq i \leq n^2$,

$$H^i(S_n) = H^{n^2-i}(S_n),$$

i.e., $H^*(S_n)$ has cohomogical dimension n^2 and satisfies Poincaré Duality.

(c) If $p - 1$ does divide n, then $H^*(S_n)$ is periodic, i.e., there is an $x \in H^{2i}(S_n)$ for some $i > 0$ such that $H^*(S_n)$ is a finitely generated free module over $\mathbf{Z}/(p)[x]$.

(d) Every sufficiently small open subgroup of S_n is cohomologically abelian in the sense that it has the same cohomology as $\mathbf{Z}_p^{n^2}$, i.e., an exterior algebra on n^2 generators.

We will describe some of the small open subgroups of S_n referred to above in 4.3.2(d). Recall (4.2.5) that S_n is the group of units in E_n that are congruent to 1 modulo the maximal ideal (S). The following result is essentially 6.3.7 of [Rav86].

Theorem 4.3.3 *Let $S_{n,i} \subset S_n$ for $i \geq 1$ be the subgroup of units in E_n congruent to 1 modulo $(S)^i$. (In particular $S_{n,1} = S_n$ and the intersection of all the $S_{n,i}$ is trivial.)*

(i) The $S_{n,i}$ are cofinal in the set of all open subgroups of S_n.

(ii) The corresponding ring of $\mathbf{F}_{p^n}$-valued functions is

$$S(n,i) = S(n)/(e_j : j < i).$$

(iii) When $i > pn/(2p-2)$, the cohomology of $S_{n,i}$ is an exterior algebra on n^2 generators .

(iv) Each $S_{n,i}$ is open and normal in S_n, and $S_{n,i}/S_{n,i+1}$ is an elementary abelian p-group of rank n. In particular the index of $S_{n,i+1}$ in S_n is p^{ni}.

The following result will be used in Chapter 8.

Theorem 4.3.4 *All finite abelian subgroups of S_n are cyclic. S_n contains an element of order p^{i+1} if and only if n is divisible by $(p - 1)p^i$. (Since S_n is a pro-p-group, it has no elements finite order prime to p.)*

Proof. Since S_n is contained in the multiplicative group of the division algebra D_n, an abelian subgroup of S_n will generate a subfield of D_n. Elements of finite order are roots of unity, so they must form a cyclic subgroup. If there is an element of order p^{i+1}, D_n must contain the field K_{i+1} obtained from $\mathbf{Q}_p$ by adjoining the $(p^{i+1})^{\text{th}}$ roots of unity. The degree of this field over $\mathbf{Q}_p$ is $(p-1)p^i$.

Now we use the fact [Rav86, 6.2.12] that D_n contains every extension of $\mathbf{Q}_p$ of degree n, and these are all the maximal subfields. Hence K_i can be embedded in D_n (i.e., S_n can have an element of order p^{i+1}) if and only if n is divisible by $(p-1)p^i$ as claimed. ∎

Chapter 5

The thick subcategory theorem

In this chapter we will derive the thick subcategory theorem (3.4.3) from the nilpotence theorem (1.4.2) with the use of some standard tools from homotopy theory, which we must introduce before we can give the proof. The proof itself is identical to the one given by Hopkins in [Hop87].

5.1 Spectra

First we have to introduce the category of spectra. These objects are similar to spaces and were invented to avoid qualifying statements (such as Definition 1.4.1) with phrases such as 'up to some suspension' and 'stably.' Since the category was introduced around 1960 [Lim60], it has taken on a life of its own, as will be seen later in this book. We will say as little about it here as we can get away with, confining more details to the Appendix (A.2). The use of the word 'spectrum' in homotopy theory has no connection with its use in analysis (the spectrum of a differential operator) or in algebraic geometry (the spectrum of a commutative ring). It also has no direct connection with the term 'spectral sequence'.

Most of the theorems in this paper that are stated in terms of spaces are really theorems about spectra that we have done our best to disguise. However we cannot keep up this act any longer.

Definition 5.1.1 *A* **spectrum** X *is a collection of spaces* $\{X_n\}$ *(defined for all large values of n) and maps* $\Sigma X_n \to X_{n+1}$. *The* **suspension spectrum** *of a space X is defined by $X_n = \Sigma^n X$ with each map being the identity. The* **sphere spectrum** S^0 *is the suspension spectrum of the space*

S^0, i.e., the nth space is S^n. The ith suspension $\Sigma^i X$ of X is defined by

$$(\Sigma^i X)_n = X_{n+i}$$

for any integer i. Thus any spectrum can be suspended or desuspended any number of times.

The homotopy groups of X are defined by

$$\pi_k(X) = \lim_{\rightarrow} \pi_{n+k}(X_n)$$

and the generalized homology $E_*(X)$ is defined by

$$E_k(X) = \lim_{\rightarrow} \overline{E}_{n+k}(X_n);$$

note that the homology groups on the right are reduced while those on the left are not. In the category of spectra there is no need to distinguish between reduced and unreduced homology.

In particular, $\pi_k(S^0)$ is the stable k-stem π_k^S of 2.2.3.

The generalized cohomology of a spectrum can be similarly defined.

A spectrum X is **connective** if its homotopy groups are bounded below, i.e., if $\pi_{-k}(X) = 0$ for $k \gg 0$. It has **finite type** if $\pi_k(X)$ is finitely generated for each k. It is **finite** if it some suspension of it is equivalent to the suspension spectrum of a finite CW-complex (A.1.1).

The homotopy groups of spectra are much more manageable than those of spaces. For example, one has

$$\pi_k(\Sigma^i E) = \pi_{k-i}(E)$$

for all k and i, and a cofibre sequence (2.3.3) of spectra leads to a long exact sequence of homotopy groups as well as the usual long exact sequence of homology groups (2.3.4).

It is surprisingly difficult to give a correct definition of a map $E \to F$ of spectra. One's first guess, namely a collection of maps $E_n \to F_n$ for $n \gg 0$ making the obvious diagrams commute, turns out to be too restrictive. While such data does give a map of spectra, there are some maps one would dearly like to have that do *not* come from any such data. However this naive definition is adequate in the case where E and F are suspension spectra of finite CW-complexes, which is all we will need for this section. A correct definition is given in A.2.5.

Next we need to discuss smash products. For spaces the definition is as follows.

Definition 5.1.2 Let X and Y be spaces equipped with base points x_0 and y_0. The **smash product** $X \wedge Y$ is the quotient of $X \times Y$ obtained by *collapsing $X \times \{y_0\} \cup \{x_0\} \times Y$ to a single point. The k-fold iterated smash product of X with itself is denoted by $X^{(k)}$. For $f: X \to Y$, $f^{(k)}$ denote the evident map from $X^{(k)}$ to $Y^{(k)}$. The map f is **smash nilpotent** if $f^{(k)}$ is null homotopic for some k.*

The k-fold suspension $\Sigma^k X$ is the same as $S^k \wedge X$. For CW-complexes X and Y there is an equivalence

$$\Sigma(X \times Y) \simeq \Sigma X \vee \Sigma Y \vee \Sigma(X \wedge Y).$$

Defining the smash product of two spectra is not as easy as one would like. If E is a suspension spectrum, then there is an obvious definition of the smash product $E \wedge F$, namely

$$(E \wedge F)_n = E_0 \wedge F_n.$$

A somewhat more flexible but still unsatisfactory definition is the following.

Definition 5.1.3 *For spectra E and F, the **naive smash product** is defined by*

$$(E \wedge F)_{2n} = E_n \wedge F_n$$
$$(E \wedge F)_{2n+1} = \Sigma E_n \wedge F_n$$

where the map

$$\Sigma E_n \wedge \Sigma F_n = \Sigma(E \wedge F)_{2n+1} \to (E \wedge F)_{2n+2} = E_{n+1} \wedge F_{n+1}$$

is the smash product of the maps $\Sigma E_n \to E_{n+1}$ and $\Sigma F_n \to F_{n+1}$.

However the correct definition of the smash product of two spectra is very difficult; we refer the interested reader to the lengthy discussion in Adams [Ada74, III.4]. In this section at least, the only smash products we need are with finite spectra, which are always suspension spectra, so the naive definition is adequate.

The nilpotence theorem can be stated in terms of smash products as follows.

Theorem 5.1.4 (Nilpotence theorem, smash product form) *Let*

$$F \xrightarrow{f} X$$

be a map of spectra where F is finite. Then f is smash nilpotent if $MU \wedge f$ (i.e., the evident map $MU \wedge F \to MU \wedge X$) is null homotopic.

Both this and 1.4.2 will be derived from a third form of the nilpotence theorem in Chapter 9. A more useful form of it for our purposes is the following, which we will prove at the end of Section 5.2.

Corollary 5.1.5 *Let W, X and Y be p-local finite spectra with $f: X \to Y$. Then $W \wedge f^{(k)}$ is null homotopic for $k \gg 0$ if $K(n)_*(W \wedge f) = 0$ for all $n \geq 0$.*

It is from this result that we will derive the thick subcategory theorem.

5.2 Spanier-Whitehead duality

Next we need to discuss Spanier-Whitehead duality, which is treated in more detail in [Ada74, III.5].

Theorem 5.2.1 *For a finite spectrum X there is a unique finite spectrum DX (the **Spanier-Whitehead dual** of X) with the following properties.*

*(i) For any spectrum Y, the graded group $[X, Y]_*$ is isomorphic to $\pi_*(DX \wedge Y)$. We say that the maps $S^n \to DX \wedge Y$ and $\Sigma^n X \to Y$ that correspond under this isomorphism are **adjoint** to each other. In particular when $Y = X$, the identity map on X is adjoint to a map $e: S^0 \to DX \wedge X$.*

(ii) This isomorphism is reflected in Morava K-theory, namely

$$\operatorname{Hom}(K(n)_*(X), K(n)_*(Y)) \cong K(n)_*(DX \wedge Y).$$

In particular for $Y = X$, $K(n)_*(e) \neq 0$ when $K(n)_*(X) \neq 0$. Similar statements hold for ordinary mod p homology.

(iii) $DDX \simeq X$.

(iv) For a homology theory E_*, there is a natural isomorphism between $E_k(X)$ and $E^{-k}(DX)$.

(v) Spanier-Whitehead duality commutes with smash products, i.e., for finite spectra X and Y, $D(X \wedge Y) = DX \wedge DY$.

The Spanier-Whitehead dual DX of a finite complex X is analogous to the linear dual $V^* = \operatorname{Hom}(V, k)$ of a finite dimensional vector space V over a field k. 5.2.1(i) is analogous to the isomorphism

$$\operatorname{Hom}(V, W) \cong V^* \otimes W$$

for any vector space W. 5.2.1(iii) is analogous to the statement that $(V^*)^* = V$ and 5.2.1(v) is analogous to the isomorphism

$$(V \otimes W)^* \cong V^* \otimes W^*.$$

The geometric idea behind Spanier-Whitehead duality is as follows. A finite spectrum X is the suspension spectrum of a finite CW-complex, which we also denote by X. The latter can always be embedded in some Euclidean space $\mathbf{R}^N$ and hence in S^N. Then DX is a suitable suspension of the suspension spectrum of the complement $S^N - X$. 5.2.1(iv) is a generalization of the classical Alexander duality theorem, which says that $H_k(X)$ is isomorphic to $H^{N-1-k}(S^N - X)$. A simple example of this is the case where $X = S^k$ and it it linearly embedded in S^N. Then its complement is homotopy equivalent to S^{N-1-k}. The Alexander duality theorem says that the complement has the same cohomology as S^{N-1-k} even when the embedding of S^k in S^N is not linear, e.g. when $k = 1$, $n = 3$ and $S^1 \subset S^3$ is knotted.

Before we can proceed with the proof of the thick subcategory theorem we need an elementary lemma about Spanier-Whitehead duality. For a finite spectrum X, let $f\colon W \to S^0$ be the map such that

$$W \xrightarrow{\ f\ } S^0 \xrightarrow{\ e\ } DX \wedge X$$

is a cofibre sequence. In the category of spectra, such maps always exist. W in this case is finite, and $C_f = DX \wedge X$.

Lemma 5.2.2 *With notation as above, there is cofibre sequence*

$$C_{f(k)} \longrightarrow C_{f(k-1)} \longrightarrow \Sigma W^{(k-1)} \wedge C_f$$

for each $k > 1$.

Proof. A standard lemma in homotopy theory says that given maps

$$X \xrightarrow{\ f\ } Y \xrightarrow{\ g\ } Z$$

there is a diagram

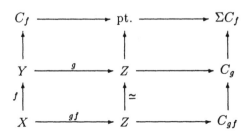

in which each row and column is a cofibre sequence. Setting $X = W^{(k)}$, $Y = W^{(k-1)}$, $Z = S^0$ and $g = f^{(k-1)}$, this diagram becomes

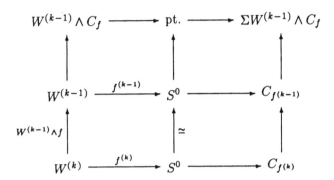

and the right hand column is the desired cofibre sequence.   ■

Proof of Corollary 5.1.5.   Let $R = DW \wedge W$ and let $e : S^0 \to R$ be the adjoint of the identity map. R is a ring spectrum (A.2.8) whose unit is e and whose multiplication is the composite

$$R \wedge R = DW \wedge W \wedge DW \wedge W \xrightarrow{DW \wedge De \wedge W} DW \wedge S^0 \wedge W = R.$$

The map $f : X \to Y$ is adjoint to $\hat{f} : S^0 \to DX \wedge Y$, and $W \wedge f$ is adjoint to the composite

$$S^0 \xrightarrow{\hat{f}} DX \wedge Y \xrightarrow{e \wedge DX \wedge Y} R \wedge DX \wedge Y = F,$$

which we denote by g. The map $W \wedge f^{(i)}$ is adjoint to the composite

$$S^0 \xrightarrow{g^{(i)}} F^{(i)} = R^{(i)} \wedge DX^{(i)} \wedge Y^{(i)} \longrightarrow R \wedge DX^{(i)} \wedge Y^{(i)},$$

the latter map being induced by the multiplication in R.

By 5.1.4 it suffices to show that $MU \wedge g^{(i)}$ is null for large i. If $x \in MU_*(F)$ is the class corresponding to the composite

$$S^0 \xrightarrow{g} F \longrightarrow MU \wedge F,$$

we need to show that

$$x^{\otimes i} \in MU_*(F^{(i)})$$

is trivial for large i.

Let n be the smallest integer so that x has a nontrivial image in $v_n^{-1} MU_*(F)$. Such an n must exist if $W \wedge f$ is essential since F is finite and p-local. Some power of each v_i for $i < n$ annihilates x, so $I_{p,n}^k x = 0$ for some k.

$K(n)_*(W \wedge f) = 0$ for all n by hypothesis, so $K(n)_*(g) = 0$ by 5.2.1(ii). This means that x has trivial image in $K(n)_*(F)$. *We claim it is divisible by* $I_{p,n}$.

Suppose this is the case. It follows that $x^{\otimes k}$ is divisible by $I_{p,n}^k$, i.e., we can write it as

$$x^{\otimes k} = \sum_i e_i y_i$$

with $e_i \in I_{p,n}^k$ and $y_i \in MU_*(F^{(k)})$. It follows that

$$
\begin{aligned}
x^{\otimes k+1} &= x \otimes \sum_i e_i y_i \\
&= \sum_i e_i x \otimes y_i \\
&= 0
\end{aligned}
$$

since $I_{p,n}^k x = 0$.

We still need to show that x is divisible by $I_{p,n}$. For this it is more convenient to use the language of BP-theory (see B.5), i.e., to replace MU by BP above and to regard x as an element of $BP_*(F)$. We need to show it is divisible by the ideal

$$I_n = (p, v_1, \cdots v_{n-1}).$$

Since x has trivial image in $K(m)_*(F)$, it is either killed by some power of v_m or is divisible by the ideal

$$J_m = (p, v_1, \cdots v_{m-1}, v_{m+1}, v_{m+2}, \cdots).$$

For $m \geq n$, Corollary 3.3.9(i) tells us that x is not v_m-torsion (since it is not v_n-torsion), so x is divisible by the ideal

$$\bigcap_{m \geq n} J_m = I_n$$

as claimed. ∎

5.3 The proof of the thick subcategory theorem

Let $\mathbf{C} \subset \mathbf{FH}_{(p)}$ be a thick subcategory. Choose n to be the smallest integer such that $\mathbf{C}$ contains a p-local finite spectrum X with $K(n)_*(X) \neq 0$. We want to show that $\mathbf{C} = \mathbf{F}_{p,n}$. It is clear from the choice of n that $\mathbf{C} \subset \mathbf{F}_{p,n}$, so it suffices to show that $\mathbf{C} \supset \mathbf{F}_{p,n}$.

Let Y be a p-local finite CW-spectrum in $\mathbf{F}_{p,n}$. From the fact that $\mathbf{C}$ is thick, it follows that $X \wedge F$ is in $\mathbf{C}$ for any finite F, so $X \wedge DX \wedge Y$ (or $C_f \wedge Y$ in the notation of 5.2.2) is in $\mathbf{C}$. Thus 5.2.2 implies that $C_{f(k)} \wedge Y$ is in $\mathbf{C}$ for all $k > 0$.

It follows from 5.2.1(ii) that $K(i)_*(f) = 0$ when $K(i)_*(X) \neq 0$, i.e., for $i \geq n$. Since $K(i)_*(Y) = 0$ for $i < n$, it follows that $K(i)_*(Y \wedge f) = 0$ for all i. Therefore by 5.1.5, $Y \wedge f^{(k)}$ is null homotopic for some $k > 0$.

Now the cofibre of a null homotopic map is equivalent to the wedge of its target and the suspension of its source, so we have

$$Y \wedge C_{f(k)} \simeq Y \vee (\Sigma Y \wedge W^{(k)}).$$

Since $\mathbf{C}$ is thick and contains $Y \wedge C_{f(k)}$, it follows that Y is in $\mathbf{C}$, so $\mathbf{C}$ contains $\mathbf{F}_{p,n}$ as desired.

Chapter 6

The periodicity theorem

In this chapter we will outline the proof of the periodicity theorem (Theorem 1.5.4). Recall that a v_n-map $f : \Sigma^d X \to X$ on a p-local finite complex X is a map such that $K(n)_*(f)$ is an isomorphism and $K(m)_*(f) = 0$ for $m \neq n$. The case $n = 0$ is uninteresting; Theorem 1.5.4 is trivial because the degree p map, which is defined for any spectrum (finite or infinite), is a v_0-map. Hence *we assume throughout this chapter that $n > 0$.*

Let $\mathbf{V}_n$ denote the collection of p-local finite spectra admitting such maps. If $K(n)_*(X) = 0$, then the trivial map is a v_n-map, so we have

$$\mathbf{V}_n \supset \mathbf{F}_{p,n+1}.$$

On the other hand, we know for algebraic reasons (3.3.11) that X cannot admit a v_n-map if $K(n-1)_*(X) \neq 0$, so

$$\mathbf{F}_{p,n} \supset \mathbf{V}_n.$$

The periodicity theorem says that $\mathbf{V}_n = \mathbf{F}_{p,n}$. The proof falls into two steps. The first is to show that $\mathbf{V}_n$ is thick; this is Theorem 6.1.5. Thus by the thick subcategory theorem, this category is either $\mathbf{F}_{p,n}$, as asserted in the periodicity theorem, or $\mathbf{F}_{p,n+1}$.

The second and harder step in the proof is to construct an example of a spectrum of type n with a v_n-map. This requires the use of the Adams spectral sequence. A brief introduction to it is given below in A.6. Its E_2-term is an Ext group for a certain module over the Steenrod algebra A. Some relevant properties of A are recalled in Section 6.2, whose main purpose is to state theorem 6.2.4. It says that a finite complex Y satisfying certain conditions always has a v_n-map. This result is proved in Section 6.3.

It then remains to construct a finite spectrum Y meeting the conditions of 6.2.4. This requires the Smith construction, which is described in Section 6.4. It is based on some work of Jeff Smith [Smi], which the author heard him lecture on in 1985 and in 1990.

6.1   Properties of v_n-maps

In this section we will prove that $\mathbf{V}_n$ is thick. We begin by observing that a self-map $f \colon \Sigma^d X \to X$ is adjoint to $\hat{f} \colon S^d \to DX \wedge X$. We will abbreviate $DX \wedge X$ by R. Now R is a ring spectrum; see A.2.8 for a definition. The unit is the map $e \colon S^0 \to DX \wedge X$ adjoint to the identity map on X (5.2.1). Since $DDX = X$ and Spanier-Whitehead duality commutes with smash products, e is dual to

$$X \wedge DX \xrightarrow{De} S^0.$$

The multiplication on R is the composite

$$DX \wedge X \wedge DX \wedge X \xrightarrow{DX \wedge De \wedge X} DX \wedge S^0 \wedge X = DX \wedge X.$$

Now we will state four lemmas, the second and fourth of which are used directly in the proof of 6.1.5. They will be proved below, and each one depends on the previous one.

Lemma 6.1.1 *For a v_n-map f as above, there is an $i > 0$ such the map induced on $K(n)_*(X)$ by f^i is multiplication by some power of v_n.*

Lemma 6.1.2 *For a v_n-map f as above, there is an $i > 0$ such that $\hat{f}^i$ is in the center of $\pi_*(R)$.*

Lemma 6.1.3 (Uniqueness of v_n-maps) *If X has two v_n-maps f and g then there are integers i and j such that $f^i = g^j$.*

Lemma 6.1.4 (Extended uniqueness) *If X and Y have v_n-maps f and g, then there are integers i and j such that the following diagram commutes for any map $h \colon X \to Y$.*

$$
\begin{array}{ccc}
\Sigma^? X & \xrightarrow{\ h\ } & \Sigma^? Y \\
\downarrow{\scriptstyle f^i} & & \downarrow{\scriptstyle g^j} \\
X & \xrightarrow{\ h\ } & Y
\end{array}
$$

Note that 6.1.3 is the special case of this where h is the identity map on X. However, we will derive 6.1.4 from 6.1.3.

Theorem 6.1.5 *The category* $\mathbf{V}_n \subset \mathbf{FH}_{(p)}$ *of finite p-local CW-spectra admitting v_n-maps is thick.*

Proof. Suppose $X \vee Y$ is in $\mathbf{V}_n$ and

$$\Sigma^d(X \vee Y) \xrightarrow{f} X \vee Y$$

is a v_n-map. By 6.1.2 we can assume that f commutes with the idempotent

$$X \vee Y \longrightarrow X \longrightarrow X \vee Y$$

and it follows that the composite

$$\Sigma^d X \longrightarrow \Sigma^d(X \vee Y) \xrightarrow{f} X \vee Y \longrightarrow X$$

is a v_n-map, so X is in $\mathbf{V}_n$.

Now suppose $h: X \to Y$ where X and Y have v_n-maps f and g. By 6.1.4 we can assume that $hf \simeq gh$, so there is a map

$$\Sigma^d C_h \xrightarrow{\ell} C_h$$

making the following diagram commute.

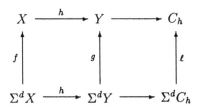

The 5-lemma implies that $K(n)_*(\ell)$ is an isomorphism.

We also need to show that $K(m)_*(\ell) = 0$ for $m \neq n$. This is *not* implied by the facts that $K(m)_*(f) = 0$ and $K(m)_*(g) = 0$. However, an easy diagram chase shows that they do imply that $K(m)_*(\ell^2) = 0$, so ℓ^2 is the desired v_n-map on C_h. It follows that C_h is in $\mathbf{V}_n$, so $\mathbf{V}_n$ is thick. ∎

Now we will give the proofs of the four lemmas stated earlier.

Proof of Lemma 6.1.1. The ring $K(n)_*(R)$ is a finite-dimensional $K(n)_*$-algebra, so the ungraded quotient $K(n)_*(R)/(v_n - 1)$ is a finite ring with a finite group of units. It follows that the group of units in $K(n)_*(R)$ itself is an extension of the group of units of $K(n)_*$ by this finite group. Therefore some power of the unit $\hat{f}_*$ is in $K(n)_*$, and the result follows. ∎

Proof of Lemma 6.1.2. Let A be a noncommutative ring, such as $\pi_*(R)$. Given $a \in A$ we define a map

$$\mathrm{ad}(a): A \longrightarrow A$$

by

$$\mathrm{ad}(a)(b) = ab - ba.$$

Thus a is in the center of A if $\mathrm{ad}(a) = 0$.

There is a formula relating $\mathrm{ad}(a^i)$ to $\mathrm{ad}^j(a)$, the jth iterate of $\mathrm{ad}(a)$, which we will prove below, namely

$$\mathrm{ad}(a^i)(x) = \sum_{j=1}^{i} \binom{i}{j} \mathrm{ad}^j(a)(x) a^{i-j}. \tag{6.1.6}$$

Now suppose $\mathrm{ad}(a)$ is nilpotent and $p^k a = 0$ for some k. We set $i = p^N$ for some large N. Then the terms on the right for large j are zero because $\mathrm{ad}(a)$ is nilpotent, and the terms for small j vanish because the binomial coefficient is divisible by p^k. Hence $\mathrm{ad}(a^i) = 0$ so a^i is in the center of A.

To apply this to the situation at hand, define

$$\Sigma^d R \xrightarrow{\ \mathrm{ad}(\hat{f})\ } R$$

to be the composite

$$S^d \wedge R \xrightarrow{\ \hat{f} \wedge R\ } R \wedge R \xrightarrow{\ 1-T\ } R \wedge R \xrightarrow{\ m\ } R$$

where T is the map that interchanges the two factors. Then for $x \in \pi_*(R)$, $\pi_*(\mathrm{ad}(\hat{f}))(x) = \mathrm{ad}(\hat{f})(x)$. By 6.1.1 (after replacing $\hat{f}$ by a suitable iterate if necessary), we can assume that $K(n)_*(f)$ is multiplication by a power of v_n, so $K(n)_*(\hat{f})$ is in the center of $K(n)_*(R)$ and $K(n)_*(\mathrm{ad}(\hat{f})) = 0$. Hence the nilpotence theorem tells us that $\mathrm{ad}(\hat{f})$ is nilpotent and the argument above applies to give the desired result.

It remains to prove (6.1.6). We have

$$\begin{aligned}
\mathrm{ad}(a^{i+1})(x) &= a^{i+1}x - xa^{i+1} \\
&= axa^i - xa^{i+1} + a^{i+1}x - axa^i \\
&= \mathrm{ad}(a)(x)a^i + a\,\mathrm{ad}(a^i)(x) \\
&= \mathrm{ad}(a)(x)a^i + \mathrm{ad}(a)(\mathrm{ad}(a^i)(x)) + \mathrm{ad}(a^i)(x)a.
\end{aligned}$$

Now we argue by induction on i, the formula being obvious for $i = 1$. From the above we have

$$\mathrm{ad}(a^{i+1})(x) \;=\; \mathrm{ad}(a)(x)a^i + \mathrm{ad}(a)\left(\sum_{j=1}^{i}\binom{i}{j}\mathrm{ad}^j(a)(x)a^{i-j}\right)$$
$$+\sum_{j=1}^{i}\binom{i}{j}\mathrm{ad}^j(a)(x)a^{i+1-j}$$

Now $\mathrm{ad}(a)$ is a derivation, i.e.,

$$\mathrm{ad}(a)(xy) = \mathrm{ad}(a)(x)y + x\,\mathrm{ad}(a)(y),$$

and it vanishes on any power of a. Hence we have

$$\mathrm{ad}(a^{i+1})(x) \;=\; \mathrm{ad}(a)(x)a^i + \sum_{j=1}^{i}\binom{i}{j}\mathrm{ad}^{j+1}(a)(x)a^{i-j}$$
$$+\sum_{j=1}^{i}\binom{i}{j}\mathrm{ad}^j(a)(x)a^{i+1-j}$$
$$=\;\sum_{j=0}^{i}\binom{i}{j}\mathrm{ad}^{j+1}(a)(x)a^{i-j}$$
$$+\sum_{j=1}^{i+1}\binom{i}{j}\mathrm{ad}^j(a)(x)a^{i+1-j}$$
$$=\;\sum_{j=1}^{i+1}\binom{i}{j-1}\mathrm{ad}^j(a)(x)a^{i+1-j}$$
$$+\sum_{j=1}^{i+1}\binom{i}{j}\mathrm{ad}^j(a)(x)a^{i+1-j}$$
$$=\;\sum_{j=1}^{i+1}\binom{i+1}{j}\mathrm{ad}^j(a)(x)a^{i+1-j}.$$

■

Proof Lemma 6.1.3.   Replacing f and g by suitable powers if necessary, we may assume that they commute with each other and that $K(m)_*(f) = K(m)_*(g)$ for all m. Hence $K(m)_*(f - g) = 0$ so $f - g$ is nilpotent. Hence there is an $i > 0$ with

$$(f - g)^{p^i} = 0.$$

Since f and g commute, we can expand this with the binomial theorem and get

$$f^{p^i} \equiv g^{p^i} \bmod p$$

from which it follows that

$$f^{p^{i+k}} \equiv g^{p^{i+k}} \bmod p^{k+1}$$

for any $k > 0$, so for sufficiently large k the two maps are homotopic. ■

Proof of Lemma 6.1.4. Let $W = DX \wedge Y$, so h is adjoint to an element $\hat{h} \in \pi_*(W)$. W has two v_n-maps, namely $DX \wedge g$ and $Df \wedge Y$, so by 6.1.3,

$$DX \wedge g^j \simeq Df^i \wedge Y$$

for suitable i and j.

Observe that W is a module spectrum over $DX \wedge X$, and the product

$$\hat{f}^i \hat{h} = (Df^i \wedge Y)\hat{h}$$

is the adjoint of hf^i. Moreover $g^j h$ is adjoint to $(DX \wedge g^j)\hat{h}$. Since these two maps are homotopic, the diagram of 6.1.4 commutes. ■

6.2   The Steenrod algebra and Margolis homology groups

Having proved 6.1.5, we need only to construct one nontrivial example of a v_n-map in order to complete the proof of the periodicity theorem. For this we will have to bring in some heavier machinery, including the Adams spectral sequence. In this section we will recall some relevant facts about the Steenrod algebra A, over which $H^*(X; \mathbf{Z}/(p))$ (for any space or spectrum X) is a module. The best reference for its properties is the classic [SE62]. This structure is crucial to what follows, as it is for most homotopy theoretic calculations.

A is a noncommutative Hopf algebra. For $p = 2$ it is generated by elements Sq^i of dimension i for $i > 0$ called **Steenrod squaring operations**. Sq^i gives a natural homomorphism

$$H^m(X) \longrightarrow H^{m+i}(X)$$

for all X and all m. The linear dual of A, A_* is easier to describe because it is a commutative algebra, namely

$$A_* = \mathbf{Z}/(2)[\xi_1, \xi_2, \cdots]$$

where the dimension of ξ_i is $2^i - 1$.

The coproduct Δ in A_* is given by

$$\Delta(\xi_n) = \sum_{0 \le i \le n} \xi_{n-i}^{2^i} \otimes \xi_i$$

where it is understood that $\xi_0 = 1$. One can show that the only primitive elements are the $\xi_1^{2^i}$, and these are dual to the generators Sq^{2^i}.

For odd primes we have

$$A_* = \mathbf{Z}/(p)[\xi_1, \xi_2, \cdots] \otimes E(\tau_0, \tau_1, \cdots)$$

with $|\xi_i| = 2p^i - 2$ and $|\tau_i| = 2p^i - 1$. The coproduct is given by

$$\Delta(\xi_n) = \sum_{0 \le i \le n} \xi_{n-i}^{p^i} \otimes \xi_i \qquad \text{and}$$

$$\Delta(\tau_n) = \tau_n \otimes 1 + \sum_{0 \le i \le n} \xi_{n-i}^{p^i} \otimes \tau_i.$$

Here the only primitives are τ_0 and $\xi_1^{p^i}$. The corresponding algebra generators of A are denoted by β (called the Bockstein operation) and $\mathcal{P}^{p^i}$.

Let $P_t^s \in A$ denote the dual (with respect to the monomial basis of A_*) to $\xi_t^{p^s}$ for $t > s \ge 0$, and (for p odd) Q_i the dual to τ_i. For $p = 2$ we will write Q_i for P_{i+1}^0. These elements and their properties have been studied by Margolis extensively in [Mar83]. They satisfy $(P_t^s)^p = 0$ and $Q_i^2 = 0$.

These conditions allow us to construct chain complexes from M in the following way. Multiplication by Q_i (and by P_t^s for $p = 2$) can be thought of as a boundary operator on M since $Q_i^2 = 0$. For P_t^s for p odd, let M_+ and M_- each be isomorphic to M and define a boundary operator d on $M_+ \oplus M_-$ as follows. For $m \in M$ let m_+ and m_- denote the corresponding elements in M_+ and M_- respectively. Then we define

$$d(m_+) = (P_t^s m)_- \qquad \text{and}$$
$$d(m_-) = ((P_t^s)^{p-1} m)_+.$$

Then $d^2 = 0$ since $(P_t^s)^p = 0$.

Definition 6.2.1 *For an A-module M, $H_*(M; Q_i)$ and $H_*(M; P_t^s)$ are the homology groups of the chain complexes defined above. These are the **Margolis homology groups** of M. If $M = H^*(X)$ we will abbreviate $H_*(M; P_t^s)$ by $H_*(X; P_t^s)$ and $H_*(M; Q_i)$ by $H_*(X; Q_i)$.*

The subalgebra of A generated by P_t^s is a truncated polynomial algebra of height p (since $(P_t^s)^p = 0$) on a single generator, while the one generated by Q_i is an exterior algebra. In both cases this subalgebra is a principal ideal domain over which any module is a direct sum of cyclic modules. There are p distinct isomorphism classes of cyclic modules (two in the Q_i case) and one sees easily that each of them other than the free one has a nontrivial Margolis homology group. Thus we have

Proposition 6.2.2 *An A-module M is free over the subalgebra generated by one of the P_t^s with $s < t$ or one of the Q_i if and only if the corresponding Margolis homology group (6.2.1) is trivial.*

The reason for our interest in these homology groups is Theorem 6.2.4 below, which combines the work of Adams-Margolis [AM71] and Anderson-Davis [AD73] ($p = 2$) or Miller-Wilkerson [MW81] (p odd). We will outline the proof in the next section.

Definition 6.2.3 *A p-local finite CW-complex Y is strongly type n if it satisfies the following conditions:*
(a) If $p = 2$, the Margolis homology groups $H_*(Y; P_t^s)$ vanish for $s + t \leq n + 1$ and $(s, t) \neq (0, n + 1)$. For $p > 2$, $H_*(Y; P_t^s)$ vanishes for $s + t \leq n$ and $H_*(Y; Q_i)$ vanishes for $i < n$.
(b) Q_n acts trivially on $H^*(Y)$.
(c) $H^*(Y)$ and $K(n)^*(Y)$ have the same rank.

Condition (b) above is actually a consequence of (c). For finite Y one can compute $K(n)^*(Y)$ from $H^*(Y)$ by means of the Atiyah-Hirzebruch spectral sequence (see theorem A.3.7 below), in which the first possibly nontrivial differential is induced by Q_n. Condition (c) is that this spectral sequence collapses, so in particular, Q_n acts trivially. The higher differentials correspond to higher order cohomology operations, so condition (c) is more subtle than the A-module structure of $H^*(X)$. Note that a strongly type n spectrum has type n (unless it is contractible) in the sense of 1.5.3 because (a) insures that $K(m)_*(X) = 0$ for $m < n$. In general a type n spectrum need not be strongly type n.

Theorem 6.2.4 *If Y is strongly type n (6.2.3) then it admits a v_n-map.*

It is easy to find finite spectra satisfying conditions (b) and (c) of 6.2.3. In particular they hold whenever the difference between the dimensions of the top and bottom cells of Y is no more than $2p^n - 2$. On the other hand condition (a) is very difficult to satisfy. Consider the following weaker condition.

Definition 6.2.5 *A finite p-local spectrum X is partially type n if it satisfies conditions (b) and (c) of 6.2.3 along with*
(a) Each of the Q_i's and P_t^0's in 6.2.3(a) acts nontrivially on $H^*(X)$.

Note that 6.2.5(a) is weaker than 6.2.3(a) because the latter requires $H^*(Y)$ to be free over the subalgebras generated by each of the elements in question. Moreover, 6.2.5(a) says nothing about the action of the P_t^s for $s > 0$. A partially type n spectrum need not have type n as in 1.5.3 because (a) does not guarantee that $K(m)_*(X) = 0$ for $m < n$.

In Section 6.4 we will describe a construction which converts a partially type n spectrum to one satisfying the conditions of 6.2.3, thereby proving the periodicity theorem. The following is an example of a partially type n spectrum.

Lemma 6.2.6 *Let* $B = B\mathbf{Z}/(p)$ *denote the classifying space for the group with p elements, B^k its k-skeleton, and $B_j^k = B^k/B^{j-1}$ (the subquotient with bottom cell in dimension j and top cell in dimension k). It can be constructed as a CW-complex with exactly one cell in each positive dimension. Then $B_2^{2p^n}$ is partially type n.*

Proof. We will assume that p is odd, leaving the case $p = 2$ as an exercise. B is the space L described in [SE62, V.5]. $B_2^{2p^n}$ satisfies condition (b) for dimensional reasons. For (c), the Atiyah-Hirzebruch spectral sequence (A.3.7) for $K(n)_*(B_2^{2p^n})$ collapses for dimensional reasons.

For condition (a) we have

$$H^*(B) = \mathbf{Z}/(p)[x] \otimes E(y)$$

with $|x| = 2$ and $|y| = 1$. It follows that a basis for $H^*(B_2^{2p^n})$ maps injectively to

$$\{x^k : 1 \le k \le p^n\} \cup \{yx^k : 1 \le k \le p^n - 1\}.$$

The actions of the Q_i and P_t^0 are given by

$$
\begin{aligned}
Q_i(x^k) &= 0 \\
Q_i(yx^k) &= x^{k+p^i} \\
P_t^0(x^k) &= kx^{k+p^t-1} \\
P_t^0(yx^k) &= kyx^{k+p^t-1}
\end{aligned}
$$

It follows that $Q_i(y)$ and $P_t^0(x)$ are each nontrivial. ∎

6.3 The Adams spectral sequence and the v_n-map on Y

Now we will outline the proof of 6.2.4, which relies heavily on the Adams spectral sequence.

Suppose
$$\Sigma^d Y \xrightarrow{\;f\;} Y$$

is a v_n-map; it is adjoint to an element $\hat{f} \in \pi_d(R)$ where $R = DY \wedge Y$.

The standard device for computing these homotopy groups is the Adams spectral sequence. This is an elaborate machine and our aim here is to say as little about it as we can get away with. A short account of it is given below in A.6 and the interested reader can learn much more in [Rav86]. Briefly, its E_2-term is

$$E_2^{s,t} = \mathrm{Ext}_A^{s,t}(H^*(R), \mathbf{Z}/(p))$$

where A is the mod p Steenrod algebra and it is understood that all cohomology groups have coefficients in $\mathbf{Z}/(p)$. The Ext group is bigraded because $H^*(R)$ itself is graded.

The differentials in the Adams spectral sequence have the form

$$E_r^{s,t} \xrightarrow{\;d_r\;} E_r^{s+r,t+r-1}.$$

Note that this raises s by r and lowers $t - s$ by one. In our situation, $E_2^{s,t}$ for a fixed value of $t - s$ vanishes for large s, so for a given (s,t), only finitely many of the d_r can be nontrivial. Thus $E_r^{s,t}$ is independent of r if r is large enough and we call this group $E_\infty^{s,t}$. The number $t - s$ corresponds to the topological dimension since $E_\infty^{s,t}$ is a subquotient of $\pi_{t-s}(R)$. Since R is a ring spectrum, the Adams spectral sequence is a spectral sequence of algebras, i.e., for each r $E_r^{*,*}$ is a bigraded algebra and each d_r is a derivation.

The multiplication in these algebras need *not* be commutative, not even up to sign. However there is a simple way to produce central elements in them. An element $g \in \pi_e(R)$ is central if it is in the image of $\alpha \in \pi_e(S^0)$ under the unit map $u: S^0 \to R$, because for any self-map $f : \Sigma^d Y \to Y$, the following diagram commutes up to sign.

$$
\begin{array}{ccc}
\Sigma^{d+e}Y & \xrightarrow{\;\Sigma^e f\;} & \Sigma^e Y \\
\Big\downarrow{\scriptstyle \Sigma^d g} & & \Big\downarrow{\scriptstyle g = \alpha \wedge Y} \\
\Sigma^d Y & \xrightarrow[\;f\;]{} & Y
\end{array}
$$

This happens because the diagram is obtained by smashing the maps f and α. The analogous algebraic statement is that elements in

$$\mathrm{Ext}_A(H^*(R), \mathbf{Z}/(p))$$

are central if they are in the image of

$$\mathrm{Ext}_A(\mathbf{Z}/(p), \mathbf{Z}/(p)),$$

which is the E_2-term of the Adams spectral sequence converging to the p-component of the stable homotopy groups of spheres.

The main result of [AM71] says that an A-module M (such as $H^*(Y)$) is free over a certain subalgebra of A if a certain set of Margolis homology groups vanish. The main result of [AD73] for $p = 2$ and [MW81] for p odd says that if M is free over this subalgebra, then certain Ext groups vanish. More precisely,

Lemma 6.3.1 *Let Y be strongly type n (6.2.3) and $R = DY \wedge Y$. Then the Adams Ext group for R has a vanishing line of slope $1/|v_n|$, i.e., there is a constant c such that the group*

$$\mathrm{Ext}_A^{s,t}(H^*(R), \mathbf{Z}/(p))$$

vanishes whenever $s > c + (t - s)/(2p^n - 2)$.

This means that if one depicts this Ext group in a chart where the vertical axis is s and the horizontal axis is $t - s$, then all the groups above a line with slope $1/(2p^n - 2)$ and vertical intercept c are zero.

For each integer $N \geq 0$ we define $A_N \subset A$ to be the subalgebra generated by the first $N + 1$ generators of A, namely by

$$\left\{ \mathrm{Sq}^1, \mathrm{Sq}^2, \cdots \mathrm{Sq}^{2^N} \right\}$$

for $p = 2$ and by

$$\left\{ \beta, \mathcal{P}^1, \mathcal{P}^p, \cdots \mathcal{P}^{p^{N-1}} \right\}$$

If $N \geq n$, then A_N contains the P_t^s of 6.2.3. One can approximate the group $\mathrm{Ext}_A(M, \mathbf{Z}/(p))$ by the groups $\mathrm{Ext}_{A_N}(M, \mathbf{Z}/(p))$ in the sense of the following result, which is essentially due to Adams [Ada66b].

Lemma 6.3.2 *If M is an A-module satisfying the conditions on $H^*(Y)$ in 6.2.3, then for each $N > n$ there is a constant $k_N > 0$ such that the map*

$$\mathrm{Ext}_A^{s,t}(M, \mathbf{Z}/(p)) \xrightarrow{\phi} \mathrm{Ext}_{A_N}^{s,t}(M, \mathbf{Z}/(p))$$

is an isomorphism when $s > (t - s)/(2p^n - 2) - k_N$, and these constants tend to infinity as N does.

The proof of this is not difficult and will be given at the end of the section.

Now consider the diagram

$$\begin{array}{ccc}
\text{Ext}_A(\mathbf{Z}/(p), \mathbf{Z}/(p)) & \xrightarrow{\ i\ } & \text{Ext}_A(H^*(R), \mathbf{Z}/(p)) \\
\phi \downarrow & & \downarrow \phi \\
\text{Ext}_{A_N}(\mathbf{Z}/(p), \mathbf{Z}/(p)) & \xrightarrow{\ i\ } & \text{Ext}_{A_N}(H^*(R), \mathbf{Z}/(p)) \\
\lambda \downarrow & & \downarrow \lambda \\
\text{Ext}_{E(Q_n)}(\mathbf{Z}/(p), \mathbf{Z}/(p)) & \xrightarrow{\ i\ } & \text{Ext}_{E(Q_n)}(H^*(R), \mathbf{Z}/(p)) \\
\cong \downarrow & & \downarrow \cong \\
P(v_n) & \xrightarrow{\ i\ } & P(v_n) \otimes H_*(R) \\
\downarrow & & \downarrow \\
K(n)_* & \xrightarrow{\ i\ } & K(n)_*(R)
\end{array}$$

The isomorphism

$$\text{Ext}_{E(Q_n)}(\mathbf{Z}/(p), \mathbf{Z}/(p)) \cong P(v_n)$$

is a standard calculation and

$$\text{Ext}_{E(Q_n)}(H^*(R), \mathbf{Z}/(p)) \cong P(v_n) \otimes H_*(R)$$

follows from the fact that Q_n acts trivially on $H^*(R)$. The fact that $H^*(R)$ and $K(n)^*(R)$ have the same rank gives us the injection

$$P(v_n) \otimes H_*(R) \longrightarrow K(n)_*(R).$$

We need the following lemma.

Lemma 6.3.3 *For each $N \geq n$ there is an integer $t > 0$ such that $v_n^t = \lambda(x)$ for some $x \in \text{Ext}_{A_N}(H^*(R), \mathbf{Z}/(p))$.*

The proof is similar to that of [Rav86, 3.4.10].

Now for any $N \geq n$, the element x provided by 6.3.3 and its image under i lie on a line of slope $1/|v_n|$ through the origin. This means that for N sufficiently large, 6.3.2 tells us that $i(x)$ is in the zone in which ϕ is an isomorphism. Hence there is an element $y \in \text{Ext}_A(H^*(R), \mathbf{Z}/(p))$ such that $\phi(y) = i(x)$. The situation is illustrated by the following picture, which represents the Ext chart for R.

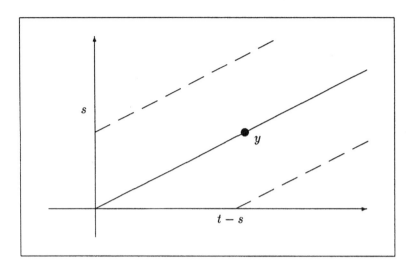

The three lines shown each have slope $1/|v_n|$. The upper one is the vanishing line given by 6.3.1. The middle one is the line through the origin, on which y lies. The bottom one is the boundary of the zone in which ϕ is an isomorphism by 6.3.2.

Lemma 6.3.4 *Some power y^{p^i} of $y \in \text{Ext}_A(H^*(R), \mathbf{Z}/(p))$ is a permanent cycle in the Adams spectral sequence for $\pi_*(R)$.*

We will prove this later in this section.
Now y^{p^i} projects to a nontrivial element in

$$K(n)_*(R) = \text{End}(K(n)_*(Y))$$

corresponding to multiplication by some power of v_n, because the same is true of

$$x^{p^i} \in \text{Ext}_{A_N}(H^*(R), \mathbf{Z}/(p)).$$

It follows that y^{p^i} corresponds to the desired v_n-map on Y. This completes our outline of the proof of 6.2.4.

Proof of Lemma 6.3.2. (Compare with [Rav86, 3.4.9]) Consider the short exact sequence

$$0 \longleftarrow M \longleftarrow A \otimes_{A_N} M \longleftarrow C \longleftarrow 0. \tag{6.3.5}$$

Now the structure of C an an A_n-module is given by

$$C \approx M \otimes \overline{A//A_N}.$$

In particular C is free over A_n and its connectivity increases exponentially with N. This means it has a vanishing line of slope $1/|v_n|$ whose vertical intercept *decreases* exponentially as N increases. We also have a change-of-rings isomorphism (see [Rav86, A1.3.13])

$$\mathrm{Ext}_A(A \otimes_{A_N} M, \mathbf{Z}/(p)) = \mathrm{Ext}_{A_N}(M, \mathbf{Z}/(p)).$$

Now we look at the long exact sequence of Ext groups associated with (6.3.5). We have

$$\mathrm{Ext}_A^{s-1}(C, \mathbf{Z}/(p))$$
$$\downarrow$$
$$\mathrm{Ext}_A^{s}(M, \mathbf{Z}/(p))$$
$$\downarrow \qquad\qquad \overset{\phi}{\nearrow}$$
$$\mathrm{Ext}_A^{s}(A \otimes_{A_N} M, \mathbf{Z}/(p)) \overset{\cong}{\longrightarrow} \mathrm{Ext}_{A_N}^{s}(M, \mathbf{Z}/(p))$$
$$\downarrow$$
$$\mathrm{Ext}_A^{s}(C, \mathbf{Z}/(p))$$

where the column is exact. From this we see that the homomorphism ϕ is an isomorphism above the vanishing line of C, which is the desired result. ∎

Proof of Lemma 6.3.4. $\phi(y) = i(x)$ is central in $\mathrm{Ext}_{A_N}(H^*(R), \mathbf{Z}/(p))$. This means that y commutes with all elements in $\mathrm{Ext}_A(H^*(R), \mathbf{Z}/(p))$ above the lower line in the picture.

Suppose that y is not a permanent cycle, i.e., that

$$d_r(y) = u \neq 0$$

for some integer r. Then u is above the line on which y lies, so it commutes with y. Since d_r is a derivation, this means that

$$d_r(y^p) = py^{p-1}u = 0,$$

i.e., y^p is a cycle in E_r. However, it could support a higher differential; we could have

$$d_{r_1}(y^p) = u_1 \neq 0$$

for some $r_1 > r$. Again, u_1 commutes with y^p and we deduce that $d_{r_1}(y^{p^2}) = 0$.

Continuing in this way, we get a sequence of integers

$$r < r_1 < r_2 < \cdots$$

with

$$d_{r_i}(y^{p^i}) = u_i.$$

For some i this differential has to be trivial because u_i must lie above the vanishing line. It follows that y^{p^i} is a permanent cycle as claimed. ∎

6.4   The Smith construction

In this section we will describe a new construction due to Jeff Smith which uses modular (characteristic p) representations of the symmetric group. It will enable us to construct a finite spectrum Y satisfying the conditions of 6.2.3.

Suppose X is a finite spectrum, and $X^{(k)}$ is its k-fold smash product. The symmetric group Σ_k acts on $X^{(k)}$ by permuting coordinates. Since we are in the stable category, it is possible to add maps, so we get an action of the group ring $\mathbf{Z}[\Sigma_k]$ on $X^{(k)}$. If X is p-local, we have an action of the p-local group ring $R = \mathbf{Z}_{(p)}[\Sigma_k]$. Now suppose e is an idempotent element ($e^2 = e$) in this group ring. Then $1 - e$ is also idempotent. For any R-module M (such as $\pi_*(X^{(k)})$) we get a splitting

$$M \cong eM \oplus (1 - e)M.$$

There is a standard construction in homotopy theory which gives a similar splitting of $X^{(k)}$ or any other spectrum on which R acts, which we write as

$$X^{(k)} \simeq eX^{(k)} \vee (1 - e)X^{(k)}.$$

In some cases one of the two summands may be trivial.

$eX^{(k)}$ can be obtained as the direct limit (see Section A.5) of the system

$$X^{(k)} \xrightarrow{\ e\ } X^{(k)} \xrightarrow{\ e\ } \cdots$$

and similarly for $(1 - e)X^{(k)}$. Since $\pi_*(X^{(k)})$ splits up in the same way, the evident map

$$X^{(k)} \longrightarrow eX^{(k)} \vee (1 - e)X^{(k)}.$$

is a homotopy equivalence.

Thus each idempotent element $e \in \mathbf{Z}_{(p)}[\Sigma_k]$ leads to a splitting of the smash product $X^{(k)}$ for any X. We can use this to convert a partially type n spectrum X (6.2.5) to a strongly type n spectrum Y (6.2.3). Actually we will use the action of Σ_k on $X^{(k\ell)}$ for a sufficiently large ℓ.

Now suppose V is a finite dimensional vector space $\mathbf{Z}/(p)$. Then $W = V^{\otimes k}$ is an R-module so we have a splitting

$$W \cong eW \oplus (1 - e)W$$

and the rank of eW is determined by that of V. There are enough idempotents e to give the following.

Theorem 6.4.1 *For each positive integer m there is an idempotent $e_m \in R$ (where the number k depends on m) such that the rank of eW above is nonzero if and only if the rank of V is at least m.*

For a proof, see C.1.5.

The example we have in mind is $V = H^*(X^{(\ell)})$ for a partially type n spectrum X. When p is odd, the action of Σ_k on $V^{\otimes k}$ is not the expected one since a minus sign must be introduced each time two odd dimensional elements are interchanged. This problem will be dealt with in C.2. In C.3 we will prove that for sufficiently large ℓ and a suitable idempotent e, the resulting spectrum is strongly type n.

Chapter 7

Bousfield localization and equivalence

In this section we will discuss localization with respect to a generalized homology theory. We attach Bousfield's name to it because the main theorem in the subject is due to him. He did invent the equivalence relation associated with it. It provides us with a very convenient language for discussing some of the concepts of this subject. A general reference for this material is [Rav84].

7.1 Basic definitions and examples

Definition 7.1.1 *Let E_* be a generalized homology theory (A.3.3). A space Y is E_*-local if whenever a map $f: X_1 \to X_2$ is such that $E_*(f)$ is an isomorphism, the map*

$$[X_1, Y] \xleftarrow{\quad f^* \quad} [X_2, Y]$$

is also an isomorphism. (For spectra, this is equivalent to the following condition: Y is E_*-local if $[X, Y]_* = 0$ whenever $E_*(X) = 0$.)

    *An E_*-**localization** of a space or spectrum X is a map η from X to an E_*-local space or spectrum X_E (which we will usually denote by $L_E X$) such that $E_*(\eta)$ is an isomorphism.*

It is easy to show that if such a localization exists, it is unique up to homotopy equivalence. The following properties are immediate consequences of the definition.

Proposition 7.1.2 *For any homology theory E_*,*

69

(i) Any inverse limit (A.5.14) of E_*-local spectra is E_*-local.
(ii) If

$$W \longrightarrow X \longrightarrow Y \longrightarrow \Sigma W$$

is a cofibre sequence and any two of W, X and Y are E_*-local, then so is the third.
(iii) If $X \vee Y$ is E_*-local, then so are X and Y.

On the other hand, a homotopy direct limit (A.5.6) of local spectra need not be local.

The main theorem in this subject, that localizations always exist, was proved by Bousfield for spaces in [Bou75] and for spectra in [Bou79b].

Theorem 7.1.3 (Bousfield localization theorem) *For any homology theory E_* and any space or spectrum X, the localization $L_E X$ of 7.1.1 exists and is functorial in X.*

The idea of the proof is the following. It is easy to see that if $L_E X$ exists, then for any map $f: X \to X'$ with $E_*(f)$ an isomorphism (such a map is called an E_*-**equivalence**), the map $\eta: X \to L_E X$ extends uniquely to X'. In other words the map η is terminal among E_*-equivalences out of X. This suggests constructing $L_E X$ as the direct limit of all such X'; this idea is due to Adams. Unfortunately it does not work because there are too many such X'; they form a class rather than a set. Bousfield found a way around these set theoretic difficulties.

If E_* is represented by a connective spectrum E (i.e., all of its homotopy groups below a given dimension are trivial), and if X is connective spectrum or a simply connected space, then the localization is relatively straightforward; it is the same as localization or completion with respect to some set of primes. The homotopy and generalized homology groups of $L_E X$ are arithmetically determined by those of X.

If either E or X fails to be connective, then $L_E X$ is far more mysterious and deserving of further study. We offer two important examples.

Example 7.1.4 *(i) X is the sphere spectrum S^0 and E_* is the homology theory associated with classical complex K-theory. $L_K S^0$ was described in [Rav84, Section 8] and it is not connective. In particular $\pi_{-2}(L_K S^0) \cong$ $\mathbf{Q/Z}$.*

(ii) Let E_* be ordinary homology H_*. Let X be a finite spectrum (such as one of the examples of 2.4.1) satisfying $K(n)_*(X) \neq 0$ with a v_n-map f (1.5.4) and let $\hat{X}$ be the telescope obtained by iterating f, i.e., $\hat{X}$ is the direct limit of the system

$$X \xrightarrow{f} \Sigma^{-d} X \xrightarrow{f} \Sigma^{-2d} X \cdots.$$

Then $L_H \hat{X}$ is contractible since $H_*(f) = 0$ and therefore $H_*(\hat{X}) = 0$. On the other hand $\hat{X}$ is not contractible since

$$K(n)_*(\hat{X}) \cong K(n)_*(X) \neq 0.$$

Lemma 7.1.5 *If E is a ring spectrum (A.2.8) then $E \wedge X$ is E_*-local for any spectrum X.*

Proof. We need to show that for any spectrum W with $E_*(W) = 0$,

$$[W, E \wedge X] = 0.$$

Given any map $f : W \to E \wedge X$, we have a diagram

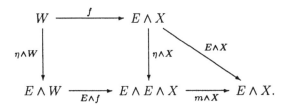

Since $E \wedge W$ is contractible, f is null. ∎

Definition 7.1.6 *For a ring spectrum E, the class of E-**nilpotent** spectra is the smallest class satisfying the following conditions.*

(i) E is E-nilpotent.

(ii) If N is E-nilpotent then so is $N \wedge X$ for any X.

(iii) The cofibre of any map between E-nilpotent spectra is E-nilpotent.

(iv) Any retract of an E-nilpotent spectrum is E-nilpotent.

*A spectrum is E-**prenilpotent** if it is E_*-equivalent to an E-nilpotent one.*

The definition of an E-nilpotent spectrum generalizes the notion of a finite Postnikov system; we replace Eilenberg-Mac Lane spectra by retracts of smash products $E \wedge X$. The following ([Bou79b, 3.8]) is an easy consequence of 7.1.5.

Proposition 7.1.7 *Every E-nilpotent spectrum is E_*-local.*

7.2   Bousfield equivalence

Recall the smash product $X \wedge Y$ was defined in 5.1.2 and the wedge $X \vee Y$ was defined in 2.1.1.

Definition 7.2.1 *Two spectra E and F are* **Bousfield equivalent** *if for each spectrum X, $E \wedge X$ is contractible if and only if $F \wedge X$ is contractible. The Bousfield equivalence class of E is denoted by $\langle E \rangle$.*

$\langle E \rangle \geq \langle F \rangle$ if for each spectrum X, the contractibility of $E \wedge X$ implies that of $F \wedge X$. We say $\langle E \rangle > \langle F \rangle$ if $\langle E \rangle \geq \langle F \rangle$ but $\langle E \rangle \neq \langle F \rangle$.

$\langle E \rangle \wedge \langle F \rangle = \langle E \wedge F \rangle$ and $\langle E \rangle \vee \langle F \rangle = \langle E \vee F \rangle$. (We leave it to the reader to verify that these classes are well defined.)

A class $\langle E \rangle$ has a **complement** *$\langle E \rangle^c$ if $\langle E \rangle \wedge \langle E \rangle^c = \langle \mathrm{pt.} \rangle$ and $\langle E \rangle \vee \langle E \rangle^c = \langle S^0 \rangle$, where S^0 is the sphere spectrum.*

The operations $\wedge$ and $\vee$ satisfy the obvious distributive laws, namely

$$(\langle X \rangle \vee \langle Y \rangle) \wedge \langle Z \rangle \quad = \quad (\langle X \rangle \wedge \langle Z \rangle) \vee (\langle Y \rangle \wedge \langle Z \rangle) \quad \text{and}$$
$$(\langle X \rangle \wedge \langle Y \rangle) \vee \langle Z \rangle \quad = \quad (\langle X \rangle \vee \langle Z \rangle) \wedge (\langle Y \rangle \vee \langle Z \rangle)$$

The following result is an immediate consequence of the definitions.

Proposition 7.2.2 *The localization functors L_E and L_F are the same if and only if $\langle E \rangle = \langle F \rangle$. If $\langle E \rangle \leq \langle F \rangle$ then $L_E L_F = L_E$ and there is a natural transformation $L_F \to L_E$.*

Notice that for any spectrum E,

$$\langle S^0 \rangle \quad \geq \quad \langle E \rangle \geq \langle \mathrm{pt.} \rangle,$$
$$\langle S^0 \rangle \wedge \langle E \rangle \quad = \quad \langle E \rangle,$$
$$\langle S^0 \rangle \vee \langle E \rangle \quad = \quad \langle S^0 \rangle,$$
$$\langle \mathrm{pt.} \rangle \vee \langle E \rangle \quad = \quad \langle E \rangle \quad \text{and}$$
$$\langle \mathrm{pt.} \rangle \wedge \langle E \rangle \quad = \quad \langle \mathrm{pt.} \rangle,$$

i.e., $\langle S^0 \rangle$ is the biggest class and $\langle \mathrm{pt.} \rangle$ is the smallest.

Not all classes have complements, and there are even classes $\langle E \rangle$ which do not satisfy

$$\langle E \rangle \wedge \langle E \rangle = \langle E \rangle. \tag{7.2.3}$$

The following definition is due to Bousfield [Bou79a].

Definition 7.2.4 **A** *is the collection of all Bousfield classes.* **DL** *(for distributive lattice) is the collection of classes satisfying (7.2.3).* **BA** *(for Boolean algebra) is the collection of classes with complements.*

Thus we have

$$\mathbf{BA} \subset \mathbf{DL} \subset \mathbf{A}$$

and both inclusions are proper. (Counterexamples illustrating this can be found in [Rav84].) If E is connective then $\langle E \rangle \in \mathbf{DL}$, and if E is a (possibly infinite) wedge of finite complexes, then $\langle E \rangle \in \mathbf{BA}$. A partial description of $\mathbf{BA}$ is given below in 7.2.9.

Let $S^0\mathbf{Q}$ denote the rational sphere spectrum, $S^0_{(p)}$ the p-local sphere spectrum, and $S^0/(p)$ the mod p Moore spectrum. Then we have

Proposition 7.2.5

$$
\begin{aligned}
\langle S^0_{(p)} \rangle &= \langle S^0\mathbf{Q} \rangle \vee \langle S^0/(p) \rangle, \\
\langle S^0\mathbf{Q} \rangle \wedge \langle S^0/(p) \rangle &= \langle \text{pt.} \rangle, \\
\langle S^0/(q) \rangle \wedge \langle S^0/(p) \rangle &= \langle \text{pt.} \rangle \quad \text{for } p \neq q, \text{ and} \\
\langle S^0 \rangle &= \langle S^0\mathbf{Q} \rangle \vee \bigvee_p \langle S^0/(p) \rangle.
\end{aligned}
$$

In particular each of these classes is in $\mathbf{BA}$.

The following result is proved in [Rav84].

Proposition 7.2.6 *(i) If*

$$W \longrightarrow X \overset{f}{\longrightarrow} Y \longrightarrow \Sigma W$$

is a cofibre sequence(2.3.3), then

$$\langle W \rangle \leq \langle X \rangle \vee \langle Y \rangle.$$

(ii) If f is smash nilpotent (5.1.2) then

$$\langle W \rangle = \langle X \rangle \vee \langle Y \rangle.$$

(iii) For a self-map $f: \Sigma^d X \to X$, let C_f denote its cofibre and let

$$\hat{X} = \varinjlim_f \Sigma^{-id} X$$

be the telescope obtained by iterating f. Then

$$
\begin{aligned}
\langle X \rangle &= \langle \hat{X} \rangle \vee \langle C_f \rangle \text{ and} \\
\langle \hat{X} \rangle \wedge \langle C_f \rangle &= \langle \text{pt.} \rangle.
\end{aligned}
$$

Two pleasant consequences of the thick subcategory theorem (3.4.3) are the following, which were the class invariance and Boolean algebra conjectures of [Rav84].

Theorem 7.2.7 (Class invariance theorem) *Let X and Y be p-local finite CW-complexes of types m and n respectively (1.5.3). Then $\langle X \rangle = \langle Y \rangle$ if and only if $m = n$, and $\langle X \rangle < \langle Y \rangle$ if and only if $m > n$.*

Proof.   Let $\mathbf{C}_X$ and $\mathbf{C}_Y$ be the smallest thick subcategories of $\mathbf{FH}_{(p)}$ containing X and Y respectively. In other words, $\mathbf{C}_X$ contains all finite complexes which can be built up from X by cofibrations and retracts. Hence each X' in $\mathbf{C}_X$ satisfies

$$\langle X' \rangle \le \langle X \rangle.$$

Since $K(m-1)_*(X) = 0$, all complexes in $\mathbf{C}_X$ are $K(m-1)_*$-acyclic, so $\mathbf{C}_X$ is contained in $\mathbf{F}_m$. On the other hand, $\mathbf{C}_X$ is not contained in $\mathbf{F}_{m+1}$ since $K(m)_*(X) \ne 0$. Hence $\mathbf{C}_X$ must be $\mathbf{F}_m$ by the thick subcategory theorem. Similarly, $\mathbf{C}_Y = \mathbf{F}_n$.

It follows that if $m = n$ then $\mathbf{C}_X = \mathbf{C}_Y$ so $\langle X \rangle = \langle Y \rangle$ as claimed. The inequalities follow similarly. $\blacksquare$

For a p-local finite CW-complex X_n of type n (1.5.3), the periodicity theorem (1.5.4) says there is a v_n-map $f \colon \Sigma^d X_n \to X_n$. We define the telescope $\hat{X}_n$ to be the direct limit of the system

$$X_n \xrightarrow{\ f\ } \Sigma^{-d} X_n \xrightarrow{\ f\ } \Sigma^{-2d} X_n \xrightarrow{\ f\ } \cdots \qquad (7.2.8)$$

Since any two choices of f agree up to iteration (6.1.3), this telescope is independent of the choice of f. Moreover, 7.2.7 implies that its Bousfield classes $\langle X_n \rangle$ and $\langle \hat{X}_n \rangle$ are independent of the choice of X_n, for a fixed n and p.

Theorem 7.2.9 (Boolean algebra theorem) *Let $\mathbf{FBA} \subset \mathbf{BA}$ be the Boolean subalgebra generated by finite spectra and their complements, and let $\mathbf{FBA}_{(p)} \subset \mathbf{FBA}$ denote the subalgebra of p-local finite spectra and their complements in $\langle S^0_{(p)} \rangle$. Then $\mathbf{FBA}_{(p)}$ the free (under complements, finite unions and finite intersections) Boolean algebra generated by the classes of the telescopes $\langle \hat{X}_n \rangle$ defined above for $n \ge 0$. In particular, the classes represented by finite spectra are*

$$\langle X_n \rangle = \bigwedge_{0 \le i < n} \langle \hat{X}_i \rangle^c.$$

In other words $\mathbf{FBA}_{(p)}$ is isomorphic to the Boolean algebra of finite and cofinite sets of natural numbers, with $\langle \hat{X}_n \rangle$ corresponding to the set $\{n\}$.

Note that this is very similar to the Boolean algebra conjecture of [Rav84, 10.8], in which $\langle \hat{X}_n \rangle$ was replaced by $\langle K(n) \rangle$. The now doubtful telescope conjecture (7.5.5) says that these two classes are the same, and 7.2.9 is phrased so that it is independent of 7.5.5.

Proof of 7.2.9. 7.2.6(iii) gives

$$\langle X_n \rangle = \langle \hat{X}_n \rangle \vee \langle X_{n+1} \rangle \quad \text{and}$$
$$\langle \hat{X}_n \rangle \wedge \langle X_{n+1} \rangle = \langle \text{pt.} \rangle.$$

This implies that

$$\langle \hat{X}_n \rangle = \langle \hat{X}_n \rangle \wedge \langle X_{n+1} \rangle^c,$$

so **FBA**$_{(p)}$ contains the indicated Boolean algebra.

On the other hand, 7.2.6(iii) also implies that

$$\langle S^0_{(p)} \rangle = \langle X_0 \rangle = \langle X_n \rangle \vee \bigvee_{0 \leq i < n} \langle \hat{X}_i \rangle,$$

from which the identification of $\langle X_n \rangle$ follows. Hence the indicated Boolean algebra contains **FBA**$_{(p)}$. ∎

7.3 The structure of $\langle MU \rangle$

The spectrum MU is described in B.2. It is known that its p-localization $MU_{(p)}$ splits into a wedge of suspensions of a 'smaller' spectrum BP, which is described in B.5. It follows that $\langle MU_{(p)} \rangle = \langle BP \rangle$ and 7.2.5 implies that

$$\langle MU \rangle = \bigvee_p \langle MU_{(p)} \rangle = \bigvee_p \langle BP \rangle$$

where the wedge on the right is over the BP's associated with the various primes p.

The class $\langle BP \rangle$ can be broken up further in terms of various spectra related to BP. A detailed account of this can be found in Section 2 of [Rav84]. The relevant spectra for our purposes are all module spectra (see A.2.8) over BP, which means that they are characterized by the structure of their homotopy groups as modules over BP_*. First we have $P(n)$ with

$$\pi_*(P(n)) = BP_*/I_n.$$

In particular $P(0)$ is BP by definition. Würgler [Wur77] has shown that each $P(n)$ is a ring spectrum. Using the construction of A.2.10, we can form the telescopes $v_n^{-1}BP$ and $v_n^{-1}P(n)$, which is denoted in the literature by $B(n)$. Closely related to these are $E(n)$ and $K(n)$ (Morava K-theory) with

$$
\begin{aligned}
E(n)_* &= \mathbf{Z}_{(p)}[v_1, v_2, \ldots v_n, v_n^{-1}] \quad \text{and} \qquad\qquad (7.3.1)\\
K(n)_* &= \mathbf{Z}/(p)[v_n, v_n^{-1}].
\end{aligned}
$$

Finally we have $H/(p)$ the mod p Eilenberg-Mac Lane spectrum representing ordinary mod p homology.

The following result was proved in [Rav84].

Theorem 7.3.2 *With notation as above,*

(a) $\langle B(n) \rangle = \langle K(n) \rangle$.

(b) $\langle v_n^{-1}BP \rangle = \langle E(n) \rangle$.

(c) $\langle P(n) \rangle = \langle K(n) \rangle \vee \langle P(n+1) \rangle$.

(d) $\langle E(n) \rangle = \bigvee_{i=0}^{n} \langle K(i) \rangle$.

(e) $\langle K(m) \rangle \wedge \langle K(n) \rangle = \langle \text{pt.} \rangle$ *for* $m \neq n$ *and* $\langle H/(p) \rangle \wedge \langle K(n) \rangle = \langle \text{pt.} \rangle$.

(f) *For* $E = K(n)$ *or* $E = H/(p)$ *and for any* X, $\langle X \rangle \wedge \langle E \rangle$ *is either* $\langle E \rangle$ *or* $\langle \text{pt.} \rangle$.

7.4   Some classes bigger than $\langle MU \rangle$

For some time after conjecturing the nilpotence theorem, we tried to prove it by showing that $\langle MU \rangle = \langle S^0 \rangle$. Eventually we disproved the latter by producing a nontrivial spectrum X with $MU_*(X) = 0$. The main tool in this construction is Brown-Comenetz duality, which was introduced in [BC76]. Their main result is the following.

Theorem 7.4.1 (Brown-Comenetz duality theorem) *Let Y be a spectrum with finite homotopy groups. Then there is a spectrum cY (the Brown-Comenetz dual of Y) such that for any spectrum X,*

$$
[X, cY]_{-i} = \text{Hom}(\pi_i(X \wedge Y), \mathbf{R}/\mathbf{Z}).
$$

In particular, $\pi_{-i}(cY) = \text{Hom}(\pi_i(Y), \mathbf{R}/\mathbf{Z})$ and $cH/(p) = H/(p)$. Moreover c is a contravariant functor on spectra with finite homotopy groups which preserves cofibre sequences and satisfies $ccY = Y$.

From this it easily follows that if $[X, cY] = 0$, then $\pi_*(X \wedge ccY) = \pi_*(X \wedge Y) = 0$. Replacing Y by cY we see that if $[X, Y] = 0$ then $\pi_*(X \wedge cY) = 0$. Now if Y is a finite complex with trivial rational homology and $X = MU$, one can show by Adams spectral sequence methods that $[X, Y] = 0$, so we conclude that

Proposition 7.4.2 *If Y is a finite complex with trivial rational homology then $MU_*(cY) = 0$.*

More details can be found in [Rav84].

The existence of a nontrivial spectrum cY with $MU_*(cY) = 0$ means that $\langle MU \rangle < \langle S^0 \rangle$.

Actually the situation is more drastic, as the following result (also proved in [Rav84]) indicates.

Theorem 7.4.3 *There are spectra $X(n)$ for $1 \leq n \leq \infty$ with $X(1) = S^0$ and $X(\infty) = MU$ such that*

$$\langle X(n) \rangle \geq \langle X(n+1) \rangle$$

for each n, with

$$\langle X(p^k - 1)_{(p)} \rangle > \langle X(p^k)_{(p)} \rangle$$

for each prime p and each $k \geq 0$.

The spectra $X(n)$ also figure in the proof of the nilpotence theorem, so we will describe them now. They are constructed in terms of vector bundles and Thom spectra. Some of the relevant background is given in B.1. Let SU denote the infinite special unitary group, i.e., the union of all the $SU(n)$'s. The Bott periodicity theorem gives us a homotopy equivalence

$$\Omega SU \longrightarrow BU$$

where BU is the classifying space of the infinite unitary group. Composing this with the loops on the inclusion of $SU(n)$ into SU, we get a map

$$\Omega SU(n) \longrightarrow BU.$$

The associated Thom spectrum (B.1.12) is $X(n)$. A routine calculation gives

$$H_*(X(n)) = \mathbf{Z}[b_1, \ldots b_{n-1}]$$

where $|b_i| = 2i$ and these generators map to generators of the same name in $H_*(MU)$ as described in B.4.1.

7.5 $E(n)$-localization and the chromatic filtration

Bousfield's theorem gives us a lot of interesting localization functors. Experience has shown that the case $E = E(n)$ (7.3.1), or equivalently (by 7.3.2(b)) $v_n^{-1}BP$, is particularly useful.

Definition 7.5.1 $L_n X$ *is* $L_{E(n)} X$ *and* $C_n X$ *denotes the fibre of the map* $X \to L_n X$.

The following result enables us to compute $BP_*(L_n X)$ in terms of $BP_*(X)$.

Theorem 7.5.2 (Localization theorem) *For any spectrum* Y,

$$BP \wedge L_n Y = Y \wedge L_n BP.$$

In particular, if $v_{n-1}^{-1} BP_*(Y) = 0$, *then*

$$BP \wedge L_n Y = Y \wedge v_n^{-1} BP,$$

i.e., $BP_*(L_n Y) = v_n^{-1} BP_*(Y)$.

The proof of this theorem and a description of $L_n BP$ will be given in Chapter 8.

Using 7.2.2 and 7.3.2(d) we get a natural transformation $L_n \to L_{n-1}$.

Definition 7.5.3 *The* **chromatic tower** *for a p-local spectrum* X *is the inverse system*

$$L_0 X \longleftarrow L_1 X \longleftarrow L_2 X \longleftarrow \cdots X.$$

The **chromatic filtration** *of* $\pi_*(X)$ *is given by the subgroups*

$$\ker (\pi_*(X) \to \pi_*(L_n X)).$$

This definition of the chromatic filtration is *not* obviously the same as the one given in 2.5.2, which was in terms of periodic maps of finite complexes. The two definitions are equivalent if the telescope conjecture 7.5.5 is true. We will refer to these as the geometric (2.5.2) and algebraic (7.5.3) definitions of the chromatic filtration.

The geometric definition is the more natural of the two. The advantage of the algebraic one is that there are methods of computing $\pi_*(L_n X)$. In particular, suppose X is a p-local finite CW-complex of type n (1.5.3) with v_n-map f. Let $\hat{X}$ be the telescope as in (7.2.8). Then $K(n)_*(f)$ is an isomorphism. The same is true of $K(i)_*(f)$ for $i < n$ since $K(i)_*(X) = 0$. Hence $E(n)_*(f)$ is an equivalence by 7.3.2(d). This means that the map $X \to L_n X$ factors uniquely through the telescope $\hat{X}$, i.e., we have a map

$$\hat{X} \xrightarrow{\lambda} L_n X. \tag{7.5.4}$$

Moreover

$$BP_*(L_n X) = v_n^{-1} BP_*(X)$$

and λ is a BP_*-equivalence.

Conjecture 7.5.5 (Telescope conjecture) *Let X be a p-local finite CW-complex of type n. Then the map λ of (7.5.4) is an equivalence.*

For $n = 0$ this statement is a triviality. The map f can be taken to be the degree p map and it is clear that $\hat{X} = L_0$ for any p-local spectrum X.

For general n it is clear that the collection of p-local type n finite complexes satisfying 7.5.5 is thick, so by the thick subcategory theorem it suffices to prove or disprove it for a single such complex. For $n = 1$, tt was proved for the mod p Moore spectrum by Mahowald [Mah82] for $p = 2$ and by Haynes Miller [Mil81] for $p > 2$. The author has recently disproved it for the type 2 complex $V(1)$ for $p \geq 5$; see [Rav92] and [Rava]. In light of this, there is no reason to think it is true for $n > 2$.

Now the v_n-torsion subgroup of $\pi_*(X)$ as defined geometrically in 2.5.2 is the kernel of the map to $\pi_*(\hat{X})$, while the corresponding subgroup defined algebraically by 7.5.3 is the kernel of the map to $\pi_*(L_n X)$. These two subgroups would be the same if the telescope conjecture were true.

What can we say when the telescope conjecture is false? The existence of the map λ of (7.5.4) means that the algebraically defined subgroup contains the geometrically defined one. However we do not know that $\pi_*(\lambda)$ is either one-to-one or onto.

The localization $L_n X$ is much better understood than the telescope $\hat{X}$. It was shown in [Rav87] that in general $\pi_*(L_n Y)$ can be computed with the Adams-Novikov spectral sequence. This is particularly pleasant in the case of a type n finite complex X. In that case there is some nice algebraic machinery for computing the E_2-term of the Adams-Novikov spectral sequence. *Indeed, it was this computability that motivated this whole program in the first place.*

We will illustrate first with the simplest possible example. Suppose our finite complex X is such that

$$BP_*(X) \cong BP_*/I_n.$$

Then $BP_*(L_n X) = v_n^{-1} BP_*/I_n$ and the E_2-term is

$$\mathrm{Ext}_{BP_*(BP)}(BP_*, v_n^{-1} BP_*/I_n).$$

This is known to be essentially the mod p continuous cohomology of the nth Morava stabilizer group S_n, described in 4.2. This isomorphism is the subject of [Rav86, Chapter 6] and a more precise statement (which would entail a distracting technical digression here) can be found there and in the change-of-rings isomorphism of B.8.8. There is also a discussion above in 4.3.

More generally, if X is a p-local finite complex of type n, then the Landweber filtration theorem (3.3.7) tells us that $BP_*(X)$ has a finite filtration in which each subquotient is a suspensions of BP_*/I_{n+i} for $i \geq 0$.

When we pass to $v_n^{-1}BP_*(X)$, we lose the subquotients with $i > 0$ and the remaining ones get converted to suspensions of $v_n^{-1}BP_*/I_n$.

Hence the Landweber filtration leads to a spectral sequence for computing the Adams-Novikov spectral sequence E_2-term,

$$\operatorname{Ext}_{BP_*(BP)}(BP_*, BP_*(L_n X)),$$

in terms of $H^*(S_n)$. It is possible to formulate its E_2-term as the cohomology of S_n with suitable twisted coefficients.

Finally, we remark that the nature of the functor L_n is partially clarified by the following.

Theorem 7.5.6 (Smash product theorem) *For any spectrum X,*

$$L_n X \cong X \wedge L_n S^0.$$

This will be proved in Chapter 8. We should point out here that in general $L_E X$ is *not* equivalent to $X \wedge L_E S^0$. Here is a simple example. Let $E = H$, the integer Eilenberg-Mac Lane spectrum. Then it is easy to show that $L_H S^0 = S^0$. On the other hand we have seen examples (7.1.4(ii)) of nontrivial Y for which L_H is contractible, so

$$L_H X \not\simeq X \wedge L_H S^0$$

in general.

The smash product theorem is a special property of the functors L_n. They may be the only localization functors with this property. The spectrum $L_1 S^0$ is well understood; its homotopy groups are given in [Rav84]. Its connective cover is essentially (precisely at odd primes) the spectrum J. $\pi_*(L_n S^0)$ is not known for any larger value of n. The computations of Shimomura-Tamura ([Shi86] and [ST86]) determine $\pi_*(L_2 V(0))$ for $p \geq 5$, where $V(0)$ denotes the mod p Moore spectrum.

A consequence of the smash product theorem is the following, which will also be proved in Chapter 8.

Theorem 7.5.7 (Chromatic convergence theorem) *For a p-local finite CW-complex X, the chromatic tower of 7.5.3 converges in the sense that*

$$X \varprojlim L_n X.$$

Chapter 8

The proofs of the localization, smash product and chromatic convergence theorems

In this chapter we will prove Theorems 7.5.2, 7.5.6 and 7.5.7. We will describe $L_n BP$ and outline the proof of the localization theorem in Section 8.1. The proof of the smash product theorem is similar in spirit to that of the periodicity theorem outlined in Chapter 6. In Section 8.2 we will explain how the thick subcategory theorem along with a theorem of Bousfield (8.2.6) can be used to reduce the problem to constructing some finite torsion free spectra with certain properties spelled out in Theorem 8.2.7. Then in Section 8.3 we will use Smith's construction along with cohomological properties of the Morava stabilizer groups to construct the required spectra. Sections 8.4 and 8.5 contain the proofs of two lemmas needed in Section 8.3. The chromatic convergence theorem (7.5.7) is proven in Section 8.6.

Apart from Section 8.1 (which is taken from [Rav87]), the material in this chapter has not appeared in print before. It is joint work with Mike Hopkins dating from 1986.

8.1 $L_n BP$ and the localization theorem

In order to describe $L_n BP$ we need to introduce the *chromatic resolution*. It is a long exact sequence of $BP_*(BP)$-comodules of the form

$$0 \longrightarrow BP_* \longrightarrow M^0 \longrightarrow M^1 \longrightarrow \cdots$$

obtained by splicing together short exact sequences

$$0 \longrightarrow N^n \longrightarrow M^n \longrightarrow N^{n+1} \longrightarrow 0.$$

These are defined in B.8. In [Rav84, Theorem 6.1] it is shown that the short exact sequence of (B.8.3) can be realized as the homotopy groups of a cofibre sequence of BP-module spectra

$$N_n BP \longrightarrow M_n BP \longrightarrow N_{n+1} BP,$$

where $N_0 BP = BP$ and $M_n BP$ is obtained from $N_n BP$ by a telescope construction which inverts v_n in homotopy.

Thus we have maps

$$BP \longleftarrow \Sigma^{-1} N_1 BP \longleftarrow \Sigma^{-2} N_2 BP \longleftarrow \cdots$$

The following result was proved as Theorem 6.2 in [Rav84].

Theorem 8.1.1 $L_n BP$ *is the cofibre of the map*

$$\Sigma^{-n-1} N_{n+1} BP \longrightarrow BP$$

described above, and there is a short exact sequence

$$0 \longrightarrow BP_* \longrightarrow \pi_*(L_n BP) \longrightarrow \Sigma^{-n} N^{n+1} \longrightarrow 0$$

which is split for $n > 0$.

To prove the localization theorem, consider the two cofibre sequences

$$
\begin{array}{ccccc}
C_n BP \wedge L_n Y & \longrightarrow & BP \wedge L_n Y & \longrightarrow L_n BP \wedge L_n Y & \quad \text{and} \\
L_n BP \wedge C_n Y & \longrightarrow & L_n BP \wedge Y & \longrightarrow L_n BP \wedge L_n Y &
\end{array}
$$

If we can show that

$$
\begin{aligned}
L_n BP \wedge C_n Y & \simeq \quad \text{pt.} \quad \text{and} & (8.1.2) \\
C_n BP \wedge L_n Y & \simeq \quad \text{pt.,} & (8.1.3)
\end{aligned}
$$

it will follow that

$$BP \wedge L_n Y \simeq L_n BP \wedge L_n Y \simeq L_n BP \wedge Y$$

as asserted in 7.5.2.

For (8.1.2), we need

Lemma 8.1.4 $\langle L_n BP \rangle = \langle v_n^{-1} BP \rangle$.

This is an easy consequence of 8.1.1 and 7.3.2; details can be found in [Rav87].

Now $C_n Y$ is $v_n^{-1} BP_*$-acyclic since it is the fibre of the map

$$Y \longrightarrow L_n Y,$$

which is a $v_n^{-1} BP_*$-equivalence. It follows that $v_n^{-1} BP \wedge C_n Y$ is contractible, so (8.1.2) follows from 8.1.4.

(8.1.3) is more difficult to prove. We know that

$$L_n Y^*(C_n BP) = [C_n BP, L_n Y]_* = 0$$

since $C_n BP$ is $v_n^{-1} BP_*$-acyclic and $L_n Y$ is $v_n^{-1} BP_*$-local. This is a cohomological statement, but we need the corresponding homological statement, namely

$$L_n Y_*(C_n BP) = \pi_*(L_n Y \wedge C_n BP) = 0.$$

Unfortunately, generalized homology is not determined by generalized cohomology except when we are computing it on a finite complex. We will need to replace $C_n BP$ by a finite complex, namely one of type $n+1$. For this reason the proof of the localization theorem requires the existence of type n complexes for all n, which was first proved by Mitchell in [Mit85]. Now it is of course a corollary of the periodicity theorem, but Mitchell's result was proved earlier.

More precisely, we need

Lemma 8.1.5 *If X is a p-local finite complex of type $n+1$, then*

$$\langle C_n BP \rangle = \langle C_n BP \wedge X \rangle.$$

Again, this is an easy consequence of 7.3.2 and the details can be found in [Rav87].

Now using Spanier-Whitehead duality (5.2.1), we have

$$\pi_*(X \wedge L_n Y) = [DX, L_n Y]_*.$$

If X has type $n+1$, it is easy to see that DX does also. Hence DX is $v_n^{-1} BP_*$-acyclic so the group above is trivial. It follows that $C_n BP \wedge X \wedge L_n Y$ is contractible, so (8.1.3) follows from 8.1.5.

This completes the proof of the main assertion of the localization theorem, namely that

$$BP \wedge L_n Y = Y \wedge L_n BP.$$

We still need to prove that when $v_{n-1}^{-1} BP_*(Y) = 0$, then

$$BP_*(L_n Y) = v_n^{-1} BP_*(Y). \tag{8.1.6}$$

In this case $L_{n-1}Y$ is contractible, so 8.1.1 gives

$$BP \wedge Y = Y \wedge \Sigma^{-n} N_n BP. \tag{8.1.7}$$

8.1.1 also gives a cofibre sequence

$$\Sigma^{-n} M_n BP \longrightarrow L_n BP \longrightarrow L_{n-1} BP.$$

Smashing this with Y gives

$$BP \wedge L_n Y = Y \wedge \Sigma^{-n} M_n BP. \tag{8.1.8}$$

Now $M_n BP$ is obtained from $N_n BP$ by a telescope construction which inverts v_n in its homotopy. Hence (8.1.7) and (8.1.8) imply that $BP_*(L_n Y)$ is obtained in a similar way from $BP_*(Y)$, thereby proving (8.1.6).

8.2 Reducing the smash product theorem to a special example

Definition 8.2.1 *A spectrum E is **smashing** if the Bousfield localization $L_E X$ is equivalent to the smash product $X \wedge L_E S^0$ for all spectra X.*

Note that since $L_E X$ depends only on the Bousfield class $\langle E \rangle$, the same is true of the question of whether E is smashing.

The following is proved in [Rav84, 1.27].

Proposition 8.2.2 *Let E be a ring spectrum and let T be $L_E S^0$. Then the following are equivalent:*

(i) E is smashing.

(ii) $\langle E \rangle = \langle T \rangle$.

(iii) Every direct limit of E_*-local spectra is E_*-local.

(iv) L_E commutes with direct limits.

Bousfield has proved the following in [Bou79b, 3.5] and [Bou79a, 2.9].

Theorem 8.2.3 *Let B be a possibly infinite wedge of finite spectra. Then* $\langle B \rangle$ *is in the Boolean algebra* **BA** *(7.2.4). If* $\langle E \rangle = \langle B \rangle^c$ *then E is smashing.*

In [Bou79b, 3.4] he conjectured that all smashing spectra arise in this way. However, the failure of the telescope conjecture for $n = 2$ means that $\langle E(2) \rangle$, which is smashing by 7.5.6, is not the expected complement of an infinite wedge of finite spectra, so Bousfield's conjecture also fails.

For any spectrum X, one has an E_*-equivalence

$$X \longrightarrow X \wedge L_E S^0.$$

The difficulty is that the target may not be local in general; it will be if X is finite.

Recall (7.1.6) that a spectrum is E-prenilpotent if its localization is E-nilpotent, i.e. if it can be built up in a finite number of stages from retracts of spectra of the form $E \wedge X$. If the sphere spectrum is E-prenilpotent, $L_E S^0$ is E-nilpotent, so $X \wedge L_E S^0$ is local for any spectrum X, so we have the following ([Bou79b, 3.9]).

Proposition 8.2.4 *A ring spectrum E is smashing if the sphere spectrum is E-prenilpotent.*

It follows immediately from the definitions that the set of p-local finite spectra which are E-prenilpotent is thick. Hence the thick subcategory theorem 3.4.3 gives us the following result, which is the starting point of our proof of 7.5.6.

Proposition 8.2.5 *A p-local ring spectrum E is smashing if there is a nontrivial E-prenilpotent p-local finite spectrum Y with torsion free homology.*

Such spectra will be constructed in the next section.

When $\pi_*(E)$ is countable (which it is in the cases of interest), Bousfield's convergence theorem A.6.11

gives a condition on the E-based Adams spectral sequence which guarantees that all spectra are E-prenilpotent and E is therefore smashing. A.6.11 is actually a characterization; a countable ring spectrum E is smashing if and only the condition is satisfied. The condition requires the existence of a horizontal vanishing line in $E_\infty(X)$ for every finite X. In the classical case ($E = H/p$) the E_∞-term for the sphere spectrum is known to have a vanishing line of positive slope, so we see again that H/p is *not* smashing.

We will need the following relative form of A.6.11.

Theorem 8.2.6 *Let E be a ring spectrum with $\pi_*(E)$ countable. Then all spectra of the form $Y \wedge W$ for a fixed Y are E-prenilpotent if and only if the E-based Adams spectral sequence satisfies the following condition: There exists a positive integer s_0 and a function φ such that for every finite spectrum X,*

$$E_\infty^{s,*}(Y \wedge X) = 0 \quad for \quad s > s_0$$

and

$$E_r^{s,*}(Y \wedge X) = E_\infty^{s,*}(Y \wedge X) \quad for \quad r > \varphi(s).$$

Corollary 8.2.7 *Let E be a ring spectrum with $\pi_*(E)$ countable. Then a spectrum Y is E-prenilpotent if the E-based Adams spectral sequence satisfies the following condition: There exist positive integers r_0 and s_0 such that for every finite spectrum X,*

$$E_{r_0}^{s,*}(Y \wedge X) = 0 \quad for \quad s > s_0,$$

i.e., there is a horizontal vanishing line of height s_0 in $E_{r_0}(Y \wedge X)$ for every finite X.

Proof.   The condition given here is stronger than that of 8.2.6 since it implies that $E_{s_0}^{s,t} = E_\infty^{s,t}$ for all s and t.   ∎

Outline of proof of Theorem 8.2.6.   We will describe how the proof of A.6.11 given in [Bou79b] can be modified to prove 8.2.6. It is observed there that the Adams spectral sequence condition follows easily from the prenilpotence assumption, and that it suffices to show that the sphere spectrum is prenilpotent. In our case it suffices for similar reasons to show that Y is E-prenilpotent.

In Bousfield's proof, various constructions are made involving the canonical Adams resolution for S^0. If we replace it with the one for Y, then a similar argument implies that Y is E-prenilpotent as claimed.   ∎

8.3   Constructing a finite torsion free prenilpotent spectrum

In this section we will construct a finite p-local torsion free spectrum Y satisfying the condition of 8.2.7 for $E = L_n BP$. The smash product theorem will follow by 8.2.5. The construction will depend on two lemmas (8.3.5 and 8.3.7) that will be proved in Sections 8.4 and 8.5.

We will need to study various Ext groups. As in (B.8.1) we will use the notation

$$\mathrm{Ext}(M) = \mathrm{Ext}_{BP_*(BP)}(BP_*, M)$$

for a $BP_*(BP)$-comodule M.

The first step in constructing the desired spectrum is the following.

Lemma 8.3.1 *A finite p-local spectrum Y with torsion free homology is $L_n BP$-prenilpotent if there is a positive integer s_0 with*

$$\mathrm{Ext}^s(v_m^{-1} BP_*(Y)/I_m) = 0$$

for $s > s_0$ and $0 \leq m \leq n$.

To prove this we need to study the Adams spectral sequence based on $L_n BP_*$ for a spectrum Y. In [Rav87, Lemma 10] it is shown that this coincides with the Adams spectral sequence for $L_n X$ based on BP_*. Hence 7.5.2 implies

Proposition 8.3.2 *A spectrum Y is $L_n BP$-prenilpotent if and only if $L_n Y$ is BP-prenilpotent.*

Proof. Since $L_n BP$ and $v_n^{-1} BP$ have the same Bousfield type, the $L_n BP$-localization of Y is $L_n Y$. Thus to say that Y is $L_n BP$-prenilpotent is to say that $L_n Y$ can be built up with a finite number of cofibrations from retracts of smash products $X \wedge L_n BP$ for various X. Now by 7.5.2,

$$X \wedge L_n BP = L_n X \wedge BP,$$

so the same construction shows that $L_n Y$ is BP-nilpotent and hence BP-prenilpotent.

Conversely, suppose that $L_n Y$ is BP-prenilpotent. It is BP-local and hence BP-nilpotent, i.e., it can be built up with a finite number of cofibrations from retracts of smash products $X \wedge BP$ for various X. If we apply L_n to everything in sight, we see that $L_n L_n Y = L_n Y$ is built up from retracts of

$$L_n(X \wedge BP),$$

so it suffices to show that this is the same as $X \wedge L_n BP$. To see this, observe that the latter is $L_n BP_*$-equivalent to $X \wedge BP$, and it is local by 7.1.5. It follows that there is a factorization

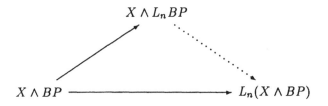

and the factorization is the desired equivalence, since both spectra are local and $L_n BP_*$-equivalent to $X \wedge BP$.   ■

Let $M_n Y$ be n^{th} suspension of the fibre of the map

$$L_n Y \longrightarrow L_{n-1} Y$$

for $n > 0$, and let $M_0 Y = L_0 Y$. (See [Rav84, 5.10] for more discussion.) We can reformulate 8.3.2 as

Proposition 8.3.3 *A spectrum Y is $L_n BP$-prenilpotent if $M_m Y$ is BP-prenilpotent for $0 \leq m \leq n$.*

Proof. $L_n Y$ is built up from the spectra $M_m Y$ by a series of cofibre sequences. It follows that if each $M_m Y$ is BP-prenilpotent, then so is $L_n Y$. The result then follows from 8.3.2.   ■

An easy consequence of the localization theorem 7.5.2 is the following.

Corollary 8.3.4 *If Y is as in 8.3.1, then*

$$BP_*(M_m Y) = BP_*(Y) \otimes M^m$$

where M^m is as in (B.8.5).

Proof of Lemma 8.3.1.   By 8.3.3 it suffices to show that $M_m Y$ is BP-prenilpotent for $0 \leq m \leq n$. Hence we need to look at the Adams-Novikov spectral sequence for $M_m Y \wedge X$ for finite X. The localization theorem (7.5.2) implies that

$$BP \wedge M_m Y \wedge X = M_m BP \wedge Y \wedge X$$

where $M_m BP$ is a BP-module spectrum with

$$\pi_*(M_m BP) = M^m.$$

Now M^m is related to $v_m^{-1} BP_*/I_m$ by a series of short exact sequences (B.8.6). In a similar way (using cofibre sequences analogous to (B.8.6)), $M_m BP$ is related to the spectrum $B(m)$, the BP-module spectrum with

$$\pi_*(B(m)) = v_m^{-1} BP/I_m.$$

It follows that

$$\text{Ext}^s(BP_*(M_m Y \wedge X)) = 0 \quad \text{for} \quad s > s_0$$

for all finite X if

$$\text{Ext}^s(v_m^{-1}BP_*(Y)/I_m) = 0 \quad \text{for} \quad s > s_0.$$

The former condition makes $M_m Y$ BP-prenilpotent by 8.2.6 and the result follows. ∎

As remarked elsewhere, the Ext groups of 8.3.1 can be computed with the help of the change-of-rings isomorphism B.8.8 and the results quoted in Section 4.3. In particular, 4.3.2(b) tells us that for $n < p - 1$, the condition of 8.3.1 holds for any Y. Hence, as indicated in [Rav87], the smash product theorem (7.5.6) for $n < p - 1$ follows from A.6.11.

For $n \geq p - 1$ it is more difficult to find a Y meeting the conditions of 8.3.1. 4.3.2(b) tells us that for any Y the condition will be satisfied for values of m not divisible by $p - 1$, but when $p - 1$ does divide m one needs some new property of Y to guarantee the condition. We know by B.8.8 that the condition of 8.3.1 is equivalent to

$$H^s(S_m; FK(m)_*(Y)) = 0 \quad \text{for} \quad 0 \leq m \leq n$$

for large s, where $FK(m)_*(Y)$ is as in (4.3.1). According to 4.3.4, S_m has a subgroup H of order p when $p - 1$ divides m. We are indebted to L. Evens for the proof of the following, which will be given in Section 8.4.

Lemma 8.3.5 *Let m be divisible by $p - 1$. If every subgroup $H \subset S_m$ of order p acts freely on the $\mathbf{F}_{p^m}$-vector space $FK(m)_*(Y)$ (i.e., if the vector space is a free module over $\mathbf{F}_{p^m}[H]$) for a finite spectrum Y, then*

$$H^s(S_m; FK(m)_*(Y)) = 0$$

for large s.

We can use Jeff Smith's methods to prove the following, which is a corollary of C.3.3.

Lemma 8.3.6 *Let m be divisible by $p - 1$ and let W be a finite torsion free spectrum such that every subgroup $H \subset S_m$ of order p acts nontrivially on $FK(m)_*(W)$. Then Σ_k acts on*

$$(W^{(p-1)})^{(k)} = W^{(k(p-1))}$$

and for some k there is an idempotent $e \in \mathbf{Z}_{(p)}[\Sigma_k]$ such that H acts freely on

$$FK(m)_*(eW^{(k(p-1))}) = eFK(m)_*(W)^{\otimes k(p-1)}.$$

It is easy to find a spectrum W meeting the requirements of 8.3.6, namely

Lemma 8.3.7 *The suspension spectrum of the complex projective space CP^j for large j satisfies the conditions for W in 8.3.6 for all $m \le n$.*

This will be proved below in Section 8.5. Combining Lemmas 8.3.5, 8.3.6 and 8.3.7 we get the finite torsion free spectrum Y needed in 8.3.1, thereby completing the proof of the smash product theorem.

8.4 Some cohomological properties of profinite groups

In this section we will give the proof of 8.3.5, which is due to L. Evens. We will use some results of Serre [Ser65] on the cohomology of profinite groups, suitably modified for our purposes. First recall the following fact.

Lemma 8.4.1 *If M is a G-module with $H^*(G; M)$ finite, then $H^*(G'; M)$ is finite for any subgroup $G' \subset G$.*

Proof. The hypotheses can be shown to imply that M has resolution of finite length by finitely generated projective G-modules. This is also a projective resolution over G', so $H^*(G'; M)$ is as claimed. ∎

Now suppose M is an $\mathbf{F}_{p^m}$-vector space and a module over a pro-p-group G and $U \subset G$ is a normal subgroup of index p, i.e., we have a group extension

$$1 \longrightarrow U \longrightarrow G \xrightarrow{\ z\ } \mathbf{Z}/(p) \longrightarrow 1. \tag{8.4.2}$$

There is a Hochschild-Serre spectral sequence [CE56, page 350] converging to $H^*(G; M)$ with

$$E_2^{s,t} = H^s(\mathbf{Z}/(p); H^t(U; M)) \quad \text{and} \quad d_r : E_r^{s,t} \longrightarrow E_r^{s+r,t-r+1}, \tag{8.4.3}$$

and it is a spectral sequence of modules over $H^*(\mathbf{Z}/(p); \mathbf{F}_p)$. The latter is

$$E(z) \otimes P(\beta(z)) \tag{8.4.4}$$

where $z \in H^1$ corresponds to the homomorphism z in (8.4.2), and β is the Bockstein operation.

The following is Proposition 5 of [Ser65] and can easily be deduced from the Hochschild-Serre spectral sequence.

Proposition 8.4.5 *Suppose U, G and M are as in (8.4.3). Suppose further that $H^*(U; M)$ is finite and $H^*(G; M)$ is infinite. Then multiplication by $\beta(z)$ (8.4.4) is an isomorphism on $H^i(G; M)$ for large i.*

In Lemma 8.3.5 we have a finite dimensional $\mathbf{F}_{p^m}$-vector space M on which the group S_m acts in such a way that every subgroup $H \subset S_m$ of order p acts freely. We want to show that $H^*(S_m; M)$ is finite. We will assume that it is infinite and derive a contradiction.

Recall the open normal subgroups of finite index $S_{m,i} \subset S_m$ of 4.3.3. Using these we can produce a sequence of open normal subgroups

$$U = U_0 \lhd U_1 \lhd \cdots U_s = S_m \tag{8.4.6}$$

such that $H^*(U; M)$ is finite and each U_i has index p in U_{i+1}. Let r be the smallest integer such that $H^*(U_r; M)$ is infinite.

Lemma 8.4.7 *With notation as above, let Σ denote the set of closed subgroups $V \subset U_r$ such that $H^*(V; M)$ is infinite. Then Σ has a minimal element.*

This is similar to Lemma 4 of [Ser65], which is stated in terms of the cohomological dimensions of the groups in question, rather than the cohomological properties of a specific module M.

The following result, which is similar in spirit to Lemma 3 of [Ser65], shows that each minimal subgroup V provided by 8.4.7 must have order p.

Lemma 8.4.8 *Let $V \subset U_r$ be a closed subgroup such that for each subgroup $U \subset V$ of index p, $H^*(U; M)$ is finite. Then either $H^*(V; M)$ is finite or V has order p.*

We will prove these two lemmas below.

Proof of Lemma 8.3.5. Let $M = FK(m)_*(Y)$ for a finite spectrum Y. This is a finite dimensional vector space over $\mathbf{F}_{p^m}$ on which S_m acts. Suppose that the lemma is false and $H^*(S_m; M)$ is infinite. Using the setup of (8.4.6), let r be the smallest integer such that $H^*(U_r; M)$ is infinite.

According to 8.4.7, there is a minimal closed subgroup $V \subset U_r$ with $H^*(V; M)$ infinite. According to 8.4.8 this subgroup must have order p. However each subgroup of order p acts freely on M by hypothesis, so its cohomology is finite. This is the desired contradiction and the result follows. ∎

Proof of Lemma 8.4.7. We can use Zorn's lemma if we know that each totally ordered subset in Σ has a lower bound in Σ. If $\{V_\lambda\}$ is such a subset, then let

$$V_0 = \bigcap_\lambda V_\lambda.$$

We need to show that $V_0 \in \Sigma$, i.e., that $H^*(V_0; M)$ is infinite.

As noted above, the group extension

$$1 \longrightarrow U_{r-1} \longrightarrow U_r \overset{z}{\longrightarrow} \mathbf{Z}/(p) \longrightarrow 1$$

induces a $H^*(\mathbf{Z}/(p); \mathbf{F}_p)$-module structure on $H^*(U_r; M)$. By 8.4.5, multiplication by $\beta(z)^k$ is nontrivial for all $k > 0$. By 8.4.1 no V_λ is contained in U_{r-1}, so we can say the same about multiplication by $\beta(z)^k$ in $H^*(V_\lambda; M)$ for each λ. Hence the same is true in

$$H^*(V_0; M) = \lim_{\longrightarrow} H^*(V_\lambda; M),$$

so $H^*(V_0; M)$ is infinite and $V_0 \in \Sigma$ as desired. ∎

Proof of Lemma 8.4.8. Let $\{z_i\}$ be a basis of $H^1(V; \mathbf{F}_p)$. This basis is finite by 4.3.2(a) and 8.4.1. Each z_i defines a homomorphism $V \to \mathbf{Z}/(p)$ whose kernel is a subgroup U of index p with $H^*(U; M)$ finite. If $H^*(V; M)$ is infinite then by 8.4.5, multiplication by each $\beta(z_i)$ is an isomorphism in large degrees. It follows that the product of all these elements in $H^*(V; \mathbf{F}_p)$ is nontrivial. According to [Ser65, Proposition 4], this means that V is elementary abelian. By 4.3.4, all finite abelian subgroups of S_m are cyclic, so V must have order p.

Alternatively (without making use of 4.3.4 or other special properties of S_m) one could argue that if the rank of V exceeds 1, then V is the direct sum of subgroups whose cohomology with coefficients in M is finite. From this one can deduce that $H^*(V; M)$ itself is finite. ∎

8.5 The action of S_m on $FK(m)_*(\mathbf{CP}^\infty)$

The purpose of this section is to prove Lemma 8.3.7 by describing the action of S_m on $FK(m)_*(\mathbf{CP}^\infty)$. We need to describe the structure of $FK(m)_*(\mathbf{CP}^\infty)$ as a comodule over $S(m)$. Recall that $FK(m)_*(\mathbf{CP}^\infty)$ has basis

$$\{b_i : i \geq 0\}$$

(see (B.1.14)) where b_i is in the image of $FK(m)_*(\mathbf{CP}^j)$ if and only if $i \leq j$. We introduce a dummy variable x and let

$$b(x) = \sum_{i \geq 0} b_i x^i.$$

The right coaction

$$FK(m)_*(CP^\infty) \xrightarrow{\psi} FK(m)_*(CP^\infty) \otimes S(m)$$

is given formally by

$$\psi(b(x)) = b(\sum_{i \geq 0} {}^{F_m} e_i x^{p^i}) \qquad (8.5.1)$$

with notation as in (4.2.6) and (4.2.7). (This is proved on page 279 of [RW77].) We obtain a formula for $\psi(b_i)$ from (8.5.1) by equating the coefficients of x^i, each of which is a finite sum.

The inclusion map $H \to S_m$ (where H has order p) induces a restriction homomorphism

$$S(m) \xrightarrow{\rho} C$$

where C is the ring of $\mathbf{F}_{p^m}$-valued functions on H. C is the linear dual of the group ring $\mathbf{F}_{p^m}[H]$. In [Rav86, 6.4.9] it was shown that its structure as a Hopf algebra is

$$C = \mathbf{F}_{p^m}[t]/(t^p - t) \quad \text{with} \quad \Delta(t) = t \otimes 1 + 1 \otimes t.$$

$M = FK(m)_*(CP^j)$ is a comodule over C where the comodule structure map ψ' is the composite

$$M \xrightarrow{\psi} M \otimes S(m) \xrightarrow{\quad M \otimes \rho \quad} M \otimes C.$$

Lemma 8.5.2 *Let $m = j(p-1)$ and let $H \subset S_m$ be any subgroup of order p. Then j is the smallest value of i such that $\rho(e_i) \in C$ is nontrivial.*

Proof. We will use the notation of (4.2.4). The subgroup H is generated by a pth root of unity $a \in E_m$, and we can write

$$a = 1 + e_1 S + e_2 S^2 + \cdots.$$

Recall that $S^m = p$ and $Se_i = e_i^p S$. We have

$$1 = a^p = (1 + e_1 S + e_2 S^2 + \cdots)^p \qquad (8.5.3)$$

We will show by induction on k that $\rho(e_k) = 0$ for $k < j$. Note that for these k, $pk < m + k$. For $k = 1$ (assuming $j > 1$), we expand (8.5.3) modulo (S^{1+p}) and get

$$\begin{aligned}
1 &\equiv (1 + e_1 S + e_2 S^2 + \cdots)^p \\
&\equiv 1 + (e_1 S)^p \\
&\equiv 1 + e_1^{1+p+p^2+\cdots p^{p-1}} S^p
\end{aligned}$$

so $e_1 = 0$.

Now we assume inductively that $e_1 = e_2 = \cdots e_{k-1} = 0$, expand (8.5.3) modulo $(S)^{pk+1}$ and get

$$
\begin{aligned}
1 &\equiv (1 + e_k S^k + e_{k+1} S^{k+1} + \cdots)^p \\
&\equiv 1 + (e_k S^k)^p \\
&\equiv 1 + e_k^{1+p^k+p^{2k}+\cdots p^{k(p-1)}} S^{pk}
\end{aligned}
$$

so $e_k = 0$ for all $k < j$.

Now we expand (8.5.3) modulo $(S)^{pj+1}$ and get

$$
\begin{aligned}
1 &\equiv (1 + e_j S^j + e_{j+1} S^{j+1} + \cdots)^p \\
&\equiv 1 + pe_j S^j + (e_j S^j)^p \\
&\equiv 1 + (e_j + e_j^{1+p^j+p^{2j}+\cdots p^{j(p-1)}}) S^{pj}
\end{aligned}
$$

so we have

$$
e_j + e_j^{1+p^j+p^{2j}+\cdots p^{j(p-1)}} = 0.
$$

This equation shows that e_j could be nontrivial. To complete the proof we need to show that $e_j = 0$ implies that $e_k = 0$ for all $k > j$, i.e., that the element a of (8.5.3) is 1.

Again we argue by induction on k. For $k > j$, $m+k < pk$, so expanding (8.5.3) modulo $(S)^{m+k+1}$ gives

$$
\begin{aligned}
1 &\equiv (1 + e_k S^k + e_{k+1} S^{k+1} + \cdots)^p \\
&\equiv 1 + pe_k S^k \\
&\equiv 1 + e_k S^{k+m},
\end{aligned}
$$

so $e_k = 0$ as desired.   ∎

Proof of Lemma 8.3.7. Let $i = m/(p-1)$; by 8.5.2, this is the smallest i such that $\rho(e_i) \in C$ is nontrivial. Then (8.5.1) gives

$$
\psi'(b_{p^i}) = b_{p^i} \otimes 1 + b_1 \otimes \rho(e_i),
$$

so H acts nontrivially on $FK(m)_*(\mathbb{C}P^j)$ for any $j \geq p^i$. By choosing j sufficiently large we can assure that this nontriviality condition holds for all m divisible by $p-1$ and $\leq n$.   ∎

8.6   Chromatic convergence

In this section we will prove the chromatic convergence theorem, 7.5.7. It says that for a p-local finite CW-complex X

$$X \simeq \varprojlim L_n X \qquad (8.6.1)$$

Recall (7.5.1) that $C_n X$ is the fibre of the localization map $X \to L_n X$, so (8.6.1) is equivalent to

$$\varprojlim \pi_i(C_n X) = \overset{1}{\varprojlim} \pi_i(C_n X) = 0 \qquad (8.6.2)$$

for each i (see A.5.15 and the preceding discussion of $\varprojlim{}^1$).

We need the following definition.

Definition 8.6.3 *A spectrum Y is E-convergent if there is a nondecreasing function $s(i)$ such that in the E-based Adams spectral sequence for Y,*

$$E_\infty^{s,s+i}(Y) = 0 \quad for \quad s > s(i).$$

This condition says that the E-based Adams spectral sequence for Y has a vanishing curve at E_∞. We know that it holds for any connective spectrum Y when $E = BP$.

Note that this condition is weaker than that of Bousfield's convergence theorem (A.6.11), which requires the function $s(i)$ to be constant.

The following two lemmas will enable us to prove the chromatic convergence theorem.

Lemma 8.6.4 *If E is connective and*

$$W \longrightarrow X \longrightarrow Y \longrightarrow \Sigma W$$

is a cofibre sequence in which W and Y are E-convergent, then X is also.

Lemma 8.6.5 *For a finite spectrum Y, the map $C_{n+1}Y \to C_n Y$ has positive BP-Adams filtration for large n.*

Proof of the chromatic convergence theorem, 7.5.7. We know that $L_n Y$ is BP-prenilpotent and hence BP-convergent by the smash product theorem(7.5.6), 8.2.4 and 8.3.2. Hence $C_n Y$ is BP-convergent by 8.6.4.

We will prove 7.5.7 by proving (8.6.2). Using 8.6.5, choose n large enough so that $C_{m+1}Y \to C_m Y$ has positive Adams filtration for each

$m \geq n$. This means that the map $C_{n+k}Y \to C_nY$ has filtration $\geq k$. Since C_nY is BP-convergent, the homomorphism

$$\pi_i(C_{n+k}Y) \longrightarrow \pi_i(C_nY)$$

is trivial for large k. This means both that the inverse limit of (8.6.2) is trivial, and that the system is Mittag-Leffler, so $\lim_{\leftarrow}{}^1$ vanishes as desired by A.5.12. ∎

Proof of Lemma 8.6.5. We need to show that the induced homomorphism

$$BP_*(C_{n+1}Y) \longrightarrow BP_*(C_nY)$$

is trivial for large n. From the localization theorem (7.5.2) we see that

$$BP \wedge C_nY = \Sigma^{-1-n}N_{n+1}BP \wedge Y,$$

so we need to look at the cofibre sequence

$$N_nBP \wedge Y \longrightarrow M_nBP \wedge Y \longrightarrow N_{n+1}BP \wedge Y \longrightarrow \Sigma N_nBP \wedge Y$$

and show that the first map is one-to-one in homotopy. This will be the case if $BP_*(Y)$ is v_n-torsion free, which it is for large n by the Landweber filtration theorem (3.3.7 and (B.5.19)). ∎

We now turn to the proof of 8.6.4. We will use the canonical Adams resolutions (A.6.1) for W, X and Y. Let i_s denote the canonical map $\overline{E}^{(s)} \to S^0$. The convergence condition on Y is equivalent to the statement that for $s > s(i)$, any composite of the form

$$F \longrightarrow \overline{E}^{(s)} \wedge Y \xrightarrow{\ i_* \wedge Y\ } Y, \qquad (8.6.6)$$

where F is a finite complex with top cell in dimension $\leq i$, is null. One could say that $i_s \wedge Y$ is 'phantom below dimension i.' Such maps have properties similar to phantom maps (maps which are null when composed with any map from a finite spectrum to the source), as the following result shows.

Lemma 8.6.7 *Let $f : X \to Y$ be phantom below dimension n as in (8.6.6). Then for any (-1)-connected spectrum W, the map $f \wedge W$ is also phantom below dimension n.*

Proof. Consider the composite

$$F \xrightarrow{e} X \wedge W \xrightarrow{f \wedge W} Y \wedge W \qquad (8.6.8)$$

where F is a finite spectrum with top cell in dimension $\leq n$. Now W is a direct limit of finite (-1)-connected spectra W_α (A.5.8). Since F is finite, e factors through $X \wedge W_\alpha$ for some α, and we can replace (8.6.8) by

$$F \xrightarrow{e} X \wedge W_\alpha \xrightarrow{f \wedge W_\alpha} Y \wedge W_\alpha.$$

The triviality of this composite is equivalent to that of the composite

$$F \wedge DW_\alpha \longrightarrow X \xrightarrow{f} Y \qquad (8.6.9)$$

where DW_α is the Spanier-Whitehead dual (5.2.1) of the finite spectrum W_α. Since W_α is (-1)-connected, $F \wedge DW_\alpha$ has top cell in dimension $\leq n$, so the composite (8.6.9) is null as desired. ∎

Proof of Lemma 8.6.4. Choose s large enough so that both $i_s \wedge W$ and $i_s \wedge Y$ are phantom below dimension i. We will prove the lemma by showing that $i_{2s} \wedge X$ is phantom below dimension i. By 8.6.7, the map $\overline{E}^{(s)} \wedge i_s \wedge Y$ is also phantom below dimension i, since $\overline{E}^{(s)}$ is (-1)-connected.

Consider the diagram

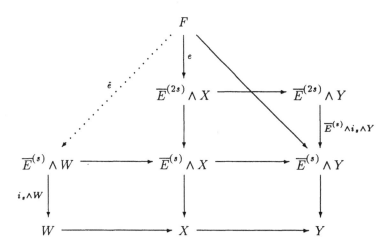

where F is a finite complex with top cell in dimension $\leq i$. Since $\overline{E}^{(s)} \wedge i_s \wedge Y$ is phantom below dimension i, the composite map $F \to \overline{E}^{(s)} \wedge Y$ is null, so the indicated lifting $\hat{e}$ exists. Its composition with $i_s \wedge W$ is null since the latter is phantom below dimension i.

It follows that the composite $(i_{2s} \wedge X)e$ is null, so $i_{2s} \wedge X$ is phantom below dimension i and X is E-convergent.   ■

Chapter 9

The proof of the nilpotence theorem

In this section we will outline the proof of the nilpotence theorem; a more detailed account is given in [DHS88]. We have previously stated it in two different guises, in terms of self-maps (1.4.2) and in terms of smash products (5.1.4). For our purposes here it is convenient to give a third statement, namely

Theorem 9.0.1 (Nilpotence theorem, ring spectrum form) *Let R be a connective ring spectrum of finite type (5.1.1 and A.2.8) and let*

$$\pi_*(R) \xrightarrow{h} MU_*(R)$$

be the Hurewicz map (A.3.4). Then $\alpha \in \pi_*(R)$ is nilpotent if $h(\alpha) = 0$.

In [DHS88] it is shown that the two previous statements are consequences of the one above. To show that 9.0.1 implies 1.4.2, let X be a finite complex and let $R = X \wedge DX$. Recall that a self-map $f: \Sigma^d X \to X$ is adjoint to a map $\hat{f}: S^d \to R$. Then $h(\hat{f})$ is nilpotent if and only if $MU_*(f)$ is.

Theorem 1.4.2 is a special case of the following statement, which is derived from 9.0.1 in [DHS88]. Suppose we have a sequence of maps of CW-spectra

$$\cdots \longrightarrow X_n \xrightarrow{f_n} X_{n+1} \xrightarrow{f_{n+1}} X_{n+2} \longrightarrow \cdots$$

with $MU_*(f_n) = 0$ for each n, and suppose there are constants $m \leq 0$ and b such that each X_n is $(mn+b)$-connected. Then the homotopy direct limit $\lim_{\to} X_n$ is contractible.

The derivation of 5.1.4 from 9.0.1 is more complicated. In the former we are given a map

$$F \xrightarrow{f} X$$

with F finite. It is adjoint to a map

$$S^0 \xrightarrow{\hat{f}} X \wedge DF$$

where DF is the Spanier-Whitehead dual (5.2.1) of F. Now f is smash nilpotent if and only if $\hat{f}$ is, and $MU \wedge f$ is null if and only if $MU \wedge \hat{f}$ is.

This means that it suffices to prove 5.1.4 for the case $F = S^0$. The hypothesis that $MU \wedge f$ is null is equivalent (since MU is a ring spectrum) to the assumption that the composite

$$S^0 \xrightarrow{f} X \longrightarrow MU \wedge X$$

is null. Since X is a homotopy direct limit of finite subspectra X_α (A.5.8), both the map f and the null homotopy for the composite above factor through some finite X_α, i.e., we have

$$S^0 \xrightarrow{f} X_\alpha \longrightarrow MU \wedge X_\alpha$$

and the composite is null.

Now let $Y = \Sigma^n X_\alpha$, where n is chosen so that Y is 0-connected. Let

$$R = \bigvee_{j \geq 0} Y^{(j)};$$

this is a connective ring spectrum of finite type with multiplication given by concatenation. Theorem 9.0.1 tells us that the element in $\pi_*(R)$ corresponding to f is nilpotent. This means that f itself is smash nilpotent, thereby proving Theorem 5.1.4.

9.1   The spectra $X(n)$

Recall the spectrum $X(n)$ of 7.4.3, the Thom spectrum associated with $\Omega SU(n)$. It is a ring spectrum so we have a Hurewicz map

$$\pi_*(R) \xrightarrow{\quad h(n) \quad} X(n)_*(R).$$

In particular $X(1) = S^0$ so $h(1)$ is the identity map. The map $X(n) \to MU$ is a homotopy equivalence through dimension $2n - 1$. It follows that if $h(\alpha) = 0$, then $h(n)(\alpha) = 0$ for large n. Hence, the nilpotence theorem will follow from

Theorem 9.1.1 *With notation as above, if $h(n+1)(\alpha) = 0$ then $h(n)(\alpha)$ is nilpotent.*

In order to prove this we need to study the spectra $X(n)$ more closely. Consider the diagram

$$
\begin{array}{ccccc}
\Omega SU(n) & \longrightarrow & \Omega SU(n+1) & \longrightarrow & \Omega S^{2n+1} \\
\big\uparrow{\scriptstyle\cong} & & \big\uparrow & & \big\uparrow \\
\Omega SU(n) & \longrightarrow & B_k & \longrightarrow & J_k S^{2n}
\end{array}
\qquad (9.1.2)
$$

in which each row is a fibration. The top row is obtained by looping the fibration

$$SU(n) \longrightarrow SU(n+1) \xrightarrow{\; e \;} S^{2n+1}$$

where e is the evaluation map which sends a matrix $m \in SU(n+1)$ to mu where $u \in \mathbf{C}^{n+1}$ is fixed unit vector.

The loop space ΩS^{2n+1} was analyzed by James [Jam55] and shown to be homotopy equivalent to a CW-complex with one cell in every dimension divisible by $2n$. $J_k S^{2n}$ denotes the kth space in the James construction on S^{2n}, which is the same thing as the $2nk$-skeleton of ΩS^{2n+1}. It can also be described as a certain quotient of the Cartesian product $(S^{2n})^k$. The space B_k is the pullback, i.e., the $\Omega SU(n)$-bundle over $J_k S^{2n}$ induced by the inclusion map into ΩS^{2n+1}.

Proposition 9.1.3 $H_*(\Omega SU(n)) = \mathbf{Z}[b_1, b_2, \ldots b_{n-1}]$ *with $|b_i| = 2i$, and*

$$H_*(B_k) \subset H_*(\Omega SU(n+1))$$

is the free module over it generated by b_n^i for $0 \leq i \leq k$.

Now the composite map

$$B_k \longrightarrow \Omega SU(n+1) \longrightarrow BU \qquad (9.1.4)$$

gives a stable bundle over B_k and we denote the Thom spectrum by F_k. Thus we have $F_0 = X(n)$ and $F_\infty = X(n+1)$. We will be especially interested in F_{p^j-1}, which we will denote by G_j. These spectra interpolate between $X(n)$ and $X(n+1)$.

The following three lemmas clearly imply 9.1.1 and hence the nilpotence theorem. Their proofs will occupy the rest of this chapter.

Lemma 9.1.5 (First lemma) *Let $\alpha^{-1}R$ be the telescope associated with $\alpha \in \pi_*(R)$ (A.2.10). If $\alpha^{-1}R \wedge X(n)$ is contractible then $h(n)_*(\alpha)$ is nilpotent.*

Lemma 9.1.6 (Second lemma) *If $h(n+1)(\alpha) = 0$ then $G_j \wedge \alpha^{-1}R$ is contractible for large j.*

The following is the hardest of the three and is the heart of the nilpotence theorem.

Lemma 9.1.7 (Third lemma) *For each $j > 0$, $\langle G_j \rangle = \langle X(n) \rangle$. In particular $\langle G_j \rangle = \langle G_{j+1} \rangle$.*

Proof of Theorem 9.1.1. We will now prove 9.1.1 assuming the three lemmas above. If $h(n+1)(\alpha) = 0$, then the telescope $\alpha^{-1}R \wedge G_j$ is contractible by 9.1.6. By 9.1.7 this means that $\alpha^{-1}R \wedge X(n)$ is also contractible. By 9.1.5, this means that $h(n)(\alpha)$ is nilpotent as claimed. ∎

9.2   The proofs of the first two lemmas

First we will prove 9.1.5. The map $\alpha: S^d \to R$ induces a self-map

$$\Sigma^d R \xrightarrow{\ \alpha\ } R.$$

The spectrum $\alpha^{-1}R \wedge X(n)$ is by definition the homotopy direct limit of

$$R \wedge X(n) \xrightarrow{\ \alpha \wedge X(n)\ } \Sigma^{-d}R \wedge X(n) \xrightarrow{\ \alpha \wedge X(n)\ } \cdots$$

It follows that each element of $X(n)_*(R)$, including $h(n)(\alpha)$, is annihilated after a finite number of steps, so $h(n)(\alpha)$ is nilpotent.

We will now outline the proof of 9.1.6. It requires the use of the Adams spectral sequence for a generalized homology theory. It is briefly introduced in A.6, and a more thorough account is given in [Rav86]. Fortunately all we require of it here is certain formal properties; we will not have to make any detailed computations.

We need to look at the Adams spectral sequence for $\pi_*(Y)$ based on $X(n+1)$-theory, for $Y = R \wedge G_j$, G_j and R. They have the following properties:

(i) The E_2-term, $E_2^{s,t}(Y)$ can be identified with a certain Ext group related to $X(n+1)$-theory, namely

$$\mathrm{Ext}_{X(n+1)_*(X(n+1))}^{s,t}(X(n+1)_*, X(n+1)_*(Y)).$$

This follows from the fact (proven in [DHS88]) that $X(n+1)$ is a flat ring spectrum (A.2.9).

(ii) $E_2^{s,t}(Y)$ vanishes unless s is nonnegative and $t - s$ exceeds the connectivity of Y

(iii) α corresponds to an element $x \in E_2^{s,s+d}(R)$ for some $s > 0$. This follows from the fact (A.6.5) that $h(n + 1)(\alpha) = 0$. The group of permanent cycle in $E_2^{0,*}(Y)$ is precisely the Hurewicz image of $\pi_*(Y)$ in $X(n + 1)_*(Y)$.

In addition we have the following property.

Lemma 9.2.1 $E_2^{s,t}(G_j)$ and $E_2^{s,t}(R \wedge G_j)$ vanish for all (s,t) above a certain line of slope

$$\frac{1}{2p^j n - 1}.$$

*(This is called a **vanishing line**.)*

We will prove this at the end of this section.

The situation is illustrated in the following picture, which is intended to illustrate $E_2^{s,t}(R \wedge G_j)$. As usual the horizontal and vertical coordinates are $t - s$ and s respectively. The powers of x all lie on a line through the origin with slope s/d. The broken line represents the vanishing line for E_2. $E_2^{s,t} = 0$ for all points (s,t) above it. For large enough j, the vanishing line has slope less than s/d and the two lines intersect as shown. It follows that x and hence $\alpha \wedge G_j$ are nilpotent, thereby proving 9.1.6.

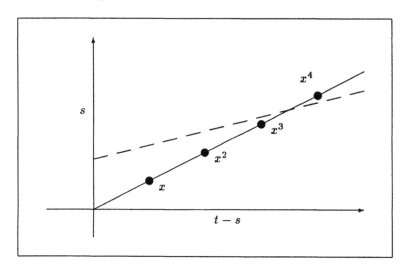

Proof of Lemma 9.2.1. We will construct a noncanonical $X(n + 1)$-based

Adams resolution for G_j, i.e. a diagram of the form

$$G_j = X_0 \xleftarrow{\;g_0\;} X_1 \xleftarrow{\;g_1\;} X_2 \longleftarrow \cdots$$

$$\left. f_0 \downarrow \qquad f_1 \downarrow \qquad f_2 \downarrow \right. \qquad (9.2.2)$$

$$K_0 \qquad\quad K_1 \qquad\quad K_2$$

as in A.6.1, such that the spectrum K_s is $(2sp^j n - s)$-connected. This will give the desired vanishing line for $E_2(G_j)$. We can get a similar resolution for $R \wedge G_j$ by smashing (9.2.2) with R, thereby proving the vanishing line for $E_2(R \wedge G_j)$.

Recall that G_j is the p-local Thom spectrum of the bundle over $B_{p^j - 1}$, which is the pullback of the fibre square

$$
\begin{array}{ccccc}
B_{p^j-1} & \xrightarrow{\;i_0\;} & \Omega SU(n+1) & \xrightarrow{\;f\;} & \Omega S^{2p^j n+1} \\
\downarrow & & \downarrow & & \cong\,\downarrow \\
J_{p^j-1}S^{2n} & \xrightarrow{\;i\;} & \Omega S^{2n+1} & \xrightarrow{\;H\;} & \Omega S^{2p^j n+1}
\end{array}
\qquad (9.2.3)
$$

The space $J_{p^j-1}S^{2n}$ is known (after localizing at p) to be fibre of the Hopf map H as shown. It follows that the same can be said of B_{p^j-1}.

The map f_0 of (9.2.2) is the Thomification of the map i_0 of (9.2.3). We will obtain the other maps f_s of (9.2.2) in a similar way. Let

$$
\begin{aligned}
Y_0 &= B_{p^j-1} \\
L_0 &= \Omega SU(n+1) \\
Y_1 &= C_{i_0}
\end{aligned}
$$

For $s \geq 0$ we will construct cofibre sequences

$$Y_s \longrightarrow L_s \longrightarrow Y_{s+1} \qquad (9.2.4)$$

which will Thomify to

$$\Sigma^s X_s \xrightarrow{\;f_s\;} \Sigma^s K_s \longrightarrow \Sigma^{s+1} X_{s+1} \qquad (9.2.5)$$

where K_s is a wedge of suspensions of $X(n+1)$ with the desired connectivity.

Our definitions of Y_s and L_s are rather longwinded. For simplicity let

$$
\begin{aligned}
X &= B_{p^j-1} \\
E &= \Omega SU(n+1) \\
B &= \Omega S^{2p^j n+1}
\end{aligned}
$$

and for $s \geq 0$ let

$$
G_s = E \times \overbrace{B \times \cdots \times B}^{s \text{ factors}}.
$$

Define maps $i_t : G_s \to G_{s+1}$ for $0 \leq t \leq s+1$ by

$$
i_t(e, b_1, b_2, \cdots b_s) = \begin{cases} (e, b_1, b_2, \cdots b_s, *) & \text{if } t = 0 \\ (e, b_1, b_2, \cdots b_t, b_t, b_{t+1}, \cdots b_s) & \text{if } 1 \leq t \leq s \\ (e, f(e), b_1, b_2, \cdots b_s) & \text{if } t = s+1. \end{cases}
$$

(The astute reader will recognize this as the cosimplicial construction associated with the Eilenberg-Moore spectral sequence, due to Larry Smith [Smi69] and Rector [Rec70].)

Then for $s \geq 1$ we define

$$
\begin{aligned}
Y_s &= G_{s-1}/\operatorname{im} i_0 \cup \operatorname{im} i_1 \cup \cdots \operatorname{im} i_{s-1} \\
L_s &= G_s/\operatorname{im} i_0 \cup \operatorname{im} i_1 \cup \cdots \operatorname{im} i_{s-1}
\end{aligned}
$$

Then for $s \geq 0$, i_s induces a map $Y_s \to L_s$ giving the cofibre sequences of (9.2.4). For $s > 0$ there are reduced homology isomorphisms

$$
\begin{aligned}
\overline{H}_*(Y_s) &= H_*(X) \otimes \overline{H}_*(B^{(s)}) \\
\overline{H}_*(L_s) &= H_*(E) \otimes \overline{H}_*(B^{(s)})
\end{aligned}
$$

This shows L_s has the desired connectivity.

Projection onto the first coordinate gives compatible maps of the G_s to E, and hence a stable vector bundle over each of them. This means that we can Thomify the entire construction. We get the cofibre sequences (9.2.5) defining the desired Adams resolution by Thomifying (9.2.4). ∎

One can also prove this result by more algebraic methods by finding a vanishing line for the corresponding Ext groups; this is the approach taken in [DHS88]. The slope one obtains is

$$
\frac{1}{p^{j+1}n - 1}
$$

which is roughly $2/p$ times the slope obtained above. In particular there is an element

$$
b_{n,j} \in \operatorname{Ext}^{2, 2p^{j+1}n}
$$

which is closely related to a self-map of G_j that will be given the same name below in (9.5.3).

All that we need to know about the slope here is that it can be made arbitrarily small by increasing n.

9.3 A paradigm for proving the third lemma

In this section we will warm up for the proof of 9.1.7 by proving a simpler result that is similar in spirit. We start with the map

$$S^q \xrightarrow{\ f\ } BU$$

(where $q = 2p - 2$) representing the generator of $\pi_q(BU) = \mathbf{Z}$. We can extend f canonically to ΩS^{2p-1}. We denote the resulting p-local Thom spectrum by $T(1)$. It has the form

$$T(1) = S^0 \cup e^q \cup e^{2q} \cup \cdots,$$

i.e., it is a p-local CW-spectrum with one cell in every q^{th} dimension. It can be shown that the qi-cell is attached to the $q(i-1)$-cell by $i\alpha_1$ where

$$\alpha_1 \in \pi_{2p-3}(S^0)$$

is the generator of the first nontrivial homotopy group of $S^0_{(p)}$ in positive dimensions.

Let

$$F_i = T(1)^{qi} \qquad \text{and} \qquad G_j = F_{p^j - 1}.$$

We use this notation to suggest the analogy with the F_i and G_j of Section 9.1.

We will show that

$$\langle G_0 \rangle = \langle G_1 \rangle. \tag{9.3.1}$$

Now $G_0 = S^0$ by definition and G_1 is the p-cell complex

$$G_1 = S^0 \cup_{\alpha_1} e^q \cup_{2\alpha_1} \cdots e^{(p-1)q}$$

There is a cofibre sequence

$$S^0 \xrightarrow{\ i\ } G_1 \xrightarrow{\ j\ } \Sigma^q F_{p-2}$$

where i is the evident inclusion map and j is the evident pinch map. Let r denote the composite

$$G_1 \xrightarrow{\ j\ } \Sigma^q F_{p-2} \xrightarrow{\ i\ } \Sigma^q G_1$$

and let K be $\Sigma^{-1}C_r$.

The following cell diagram illustrates this for $p = 3$. Each small circle represents a cell in the indicated spectrum and the numbers to the left indicate the dimensions of the various cells. The maps j and i induce homology isomorphisms in the middle dimensions, so the cofibre of r has cells only in dimensions 1 and 12.

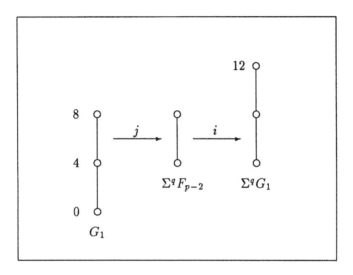

Thus we have a cofibre sequence

$$G_1 \xrightarrow{r} \Sigma^q G_1 \longrightarrow \Sigma K, \tag{9.3.2}$$

which shows that

$$\langle G_1 \rangle \geq \langle K \rangle \tag{9.3.3}$$

On the other hand, K is a 2-cell complex, i.e., there is a cofibre sequence

$$S^{pq-2} \xrightarrow{\beta_1} S^0 \longrightarrow K \tag{9.3.4}$$

where β_1 is the generator of the $(pq - 2)$-stem. It is known to be nilpotent, so 7.2.6(ii) gives

$$\langle K \rangle = \langle S^0 \rangle = \langle G_1 \rangle.$$

This along with (9.3.3) proves (9.3.1).

9.4   The Snaith splitting of $\Omega^2 S^{2m+1}$

In this section we will recall some results of Snaith [Sna74] concerning the homotopy type of $\Omega^2 S^{2m+1}$. This will be needed to prove 9.1.7 and the nilpotence theorem.

We begin by recalling the mod p homology of $\Omega^2 S^{2m+1}$.

Proposition 9.4.1 *For any $m > 0$ and any odd prime p,*

$$
\begin{aligned}
H_*(\Omega^2 S^{2m+1}; \mathbf{Z}/(p)) \;=\; & E(x_{2m-1}, x_{2pm-1}, x_{2p^2m-1}, \cdots) \\
& \otimes P(y_{2pm-2}, y_{2p^2m-2}, \cdots)
\end{aligned}
$$

where the subscript of each generator indicates its dimension. For $p = 2$,

$$
H_*(\Omega^2 S^{2m+1}; \mathbf{Z}/(2)) = P(x_{2m-1}, x_{4m-1}, x_{8m-1}, \cdots).
$$

We can also describe the action of the Steenrod algebra on this homology. Recall that for any space X, $H^*(X; \mathbf{Z}/(p))$ is a left A-module, where A denote the mod p Steenrod algebra. It follows by duality that $H_*(X; \mathbf{Z}/(p))$ is a right A-module, and the Steenrod operations *lower* dimensions instead of raising them. In particular, we have

$$
H_n(X; \mathbf{Z}/(2)) \xrightarrow{\quad Sq^i \quad} H_{n-i}(X; \mathbf{Z}/(2))
$$

$$
H_n(X; \mathbf{Z}/(p)) \xrightarrow{\quad \beta \quad} H_{n-1}(X; \mathbf{Z}/(p))
$$

$$
H_n(X; \mathbf{Z}/(p)) \xrightarrow{\quad \mathcal{P}^i \quad} H_{n-i(2p-2)}(X; \mathbf{Z}/(p))
$$

In the case of the double loop space, this action is described by

Proposition 9.4.2 *For $p > 2$,*

$$
\begin{aligned}
(x_{2p^{i+1}m-1})\beta &= y_{2p^{i+1}m-2} & \text{for } \; i \geq 0 \\
(y_{2p^{i+1}m-2})\beta &= 0 & \text{for } \; i \geq 0 \\
(x_{2p^i m-1})\mathcal{P}^j &= 0 & \text{for } \; i, j \geq 0 \\
(y_{2p^{i+1}m-1})\mathcal{P}^1 &= y_{2p^i m-2}^p & \text{for } \; i \geq 0 \\
(y_{2p^i m-1})\mathcal{P}^j &= 0 & \text{for } \; i \geq 0, j > 1
\end{aligned}
$$

For $p = 2$,

$$
\begin{aligned}
(x_{2^{i+1}m-1})Sq^1 &= y_{2^i m-1}^1 & \text{for } \; i \geq 0 \\
(x_{2^{i+1}m-1})Sq^j &= 0 & \text{for } \; i \geq 0, j > 1.
\end{aligned}
$$

Snaith proved that the suspension spectrum $\Omega^2 S^{2m+1}$ is equivalent to an infinite wedge of finite complexes which he described explicitly, i.e., he gave a decomposition of the form

$$\Sigma^\infty \Omega^2 S^{2m+1}_+ \simeq \bigvee_{i \geq 0} D_{m,i} \qquad (9.4.3)$$

for finite $D_{m,i}$. (X_+ indicates a the space X disjoint basepoint added, making $H_*(\Sigma^\infty X_+)$ isomorphic to the *unreduced* homology of X.) The $D_{m,i}$ are independent of m up to suspension.

In order to describe this decomposition in homology, we assign a weight to each generator of $H_*(\Omega^2 S^{2m+1})$ by defining

$$|x_{2p^i m-1}| = |y_{2p^i m-2}| = p^i. \qquad (9.4.4)$$

Theorem 9.4.5 (Snaith splitting theorem) *The suspension spectrum*

$$\Sigma^\infty \Omega^2 S^{2m+1}$$

has a decomposition as in (9.4.3) where $H_*(D_{m,i}; \mathbf{Z}/(p))$ is the vector space spanned by the monomials of weight i.

From now on, we assume that everything in sight has been localized at p. Inspection of (9.4.4) shows that every generator has weight divisible by p except x_{2m-1}. It follows that $D_{m,i}$ is contractible unless i is congruent to 0 or 1 mod p. We also see that

$$\begin{aligned} D_{m,0} &= S^0, \\ D_{m,1} &= S^{2m-1}, \end{aligned}$$

$D_{m,pi}$ is $(2i(pm-1)-1)$-connected, and

$$H_*(D_{m,pi+1}) = \Sigma^{2m-1} H_*(D_{m,pi}),$$

which suggests that

$$D_{m,pi+1} = \Sigma^{2m-1} D_{m,pi}. \qquad (9.4.6)$$

This is indeed the case, as one sees in the following way. $\Omega^2 S^{2m+1}$ is an H-space, which means there is a map

$$\Omega^2 S^{2m+1} \times \Omega^2 S^{2m+1} \xrightarrow{\lambda} \Omega^2 S^{2m+1}$$

with certain properties. Stably this induces maps

$$D_{m,i} \wedge D_{m,j} \xrightarrow{\lambda} D_{m,i+j}. \qquad (9.4.7)$$

In particular we have

$$\Sigma^{2m-1} D_{m,pi} = D_{m,1} \wedge D_{m,pi} \xrightarrow{\lambda} D_{m,pi+1}$$

inducing multiplication by x_{2m-1} in homology, thereby proving (9.4.6).

As remarked above, the complexes $D_{m,i}$ are independent of m up to suspension. Snaith's theorem can be reformulated as follows.

Theorem 9.4.8 *For each $m > 0$ and any prime p,*

$$\Sigma^\infty \Omega^2 S_+^{2m+1} \cong (S^0 \vee S^{2m-1}) \wedge \bigvee_{i \geq 0} \Sigma^{2i(pm-1)} D_i$$

where

$$D_i = \Sigma^{-2i(pm-1)} D_{m,pi}$$

where each D_i is a finite (-1)-connected spectrum.

In particular,

$$\begin{aligned} D_0 &= S^0 \\ D_1 &= S^0 \cup_p e^1, \end{aligned}$$

i.e., D_1 is the mod p Moore spectrum. Using the map λ of (9.4.7) we get

$$D_i \longrightarrow D_1 \wedge D_i \xrightarrow{\lambda} D_{i+1}. \qquad (9.4.9)$$

We denote this map by ℓ. It induces multiplication by y_{2pm-2} in homology.

The following result, which is originally due to Mahowald [Mah77], is at first glance somewhat surprising.

Theorem 9.4.10 *The homotopy direct limit (A.5.6)*

$$\varinjlim_i D_i$$

is the mod p Eilenberg-MacLane spectrum H/p.

This is actually rather easy to prove. We have all the tools necessary to compute the homology of this limit. We find that it can be identified with the subspace of weight 0 in

$$y_{2pm-2}^{-1} H_*(\Omega^2 S^{2m+1}),$$

which is

$$P(y_{2pm-2}^{-p^{i-1}} y_{2p^i m-2} : i > 0) \otimes E(y_{2pm-2}^{-p^{i-1}} x_{2p^i m-1} : i \geq 0).$$

As a ring this is isomorphic to the dual Steenrod algebra A_* (see B.3.4) and from 9.4.2 we can see that the right action of A on is the same as in $H_*(H/p)$.

9.5   The proof of the third lemma

We will now prove 9.1.7 using methods similar to that of Section 9.3. We need to show that $\langle G_j \rangle = \langle G_{j+1} \rangle$. Recall that $G_j = F_{p^j-1}$, and $H_*(F_k)$ is the free module over $H_*(X(n))$ generated by b_n^i for $0 \leq i \leq k$. One has inclusion maps

$$X(n) = F_0 \hookrightarrow F_1 \hookrightarrow F_2 \hookrightarrow \cdots$$

with cofibre sequences

$$F_{k-1} \longrightarrow F_k \longrightarrow \Sigma^{2kn} X(n).$$

From this it follows immediately that

$$\langle F_k \rangle \leq \langle X(n) \rangle$$

for all $k \geq 0$.

It can also be shown that (after localizing at p) there is a cofibre sequence

$$F_{kp^j-1} \longrightarrow F_{(k+1)p^j-1} \longrightarrow \Sigma^{2nkp^j} G_j.$$

In particular we have

$$G_j = F_{p^j-1} \hookrightarrow F_{2p^j-1} \hookrightarrow \cdots F_{(p-1)p^j-1} \hookrightarrow F_{p^{j+1}-1} = G_{j+1}$$

where the cofibre of each map is a suspension of G_j. This shows that

$$\langle G_j \rangle \geq \langle G_{j+1} \rangle. \tag{9.5.1}$$

It is also straightforward to show that there is a cofibre sequence

$$G_j \longrightarrow G_{j+1} \longrightarrow \Sigma^{2np^j} F_{(p-1)p^j-1}$$

which induces a short exact sequence in homology. Thus we can form the composite map

$$G_{j+1} \longrightarrow \Sigma^{2np^j} F_{(p-1)p^j-1} \longrightarrow \Sigma^{2np^j} G_{j+1}$$

in which the first map is surjective in homology while the second is monomorphic. We denote this map by $r_{n,j}$. *It is analogous to the composite shown in the cell diagram on page 107. Each cell there should be replaced by a copy of G_j.*

Then there are cofibre sequences

$$G_{j+1} \xrightarrow{\;\;r_{n,j}\;\;} \Sigma^{2np^j} G_{j+1} \longrightarrow K_{n,j} \qquad (9.5.2)$$

analogous to (9.3.2) and

$$\Sigma^{2np^{j+1}-2} G_j \xrightarrow{\;\;b_{n,j}\;\;} G_j \longrightarrow K_{n,j} \qquad (9.5.3)$$

analogous to (9.3.4). The first of these shows that

$$\langle G_{j+1} \rangle \geq \langle K_{n,j} \rangle. \qquad (9.5.4)$$

Using 7.2.6(iii), we see that *if the telescope $b_{n,j}^{-1} G_j$ is contractible* then we will have

$$\langle K_{n,j} \rangle = \langle G_j \rangle \qquad \text{so}$$
$$\langle G_{j+1} \rangle = \langle G_j \rangle \qquad \text{by (9.5.4) and (9.5.1)}$$

Thus we have reduced the nilpotence theorem to the following.

Lemma 9.5.5 *Let*

$$\Sigma^{2np^{j+1}-2} G_j \xrightarrow{\;\;b_{n,j}\;\;} G_j$$

be the map of (9.5.3). It has a contractible telescope for each n and j.

This is equivalent to the statement that for each finite skeleton of G_j, there is an iterate of $b_{n,j}$ whose restriction to the skeleton is null.
Proof. We need to look again at (9.1.2) for $k = p^j - 1$. The map

$$J_{p^j-1} S^{2n} \longrightarrow \Omega S^{2n+1}$$

is known (after localizing at p) to the be inclusion of the fibre of a map

$$\Omega S^{2n+1} \xrightarrow{H} \Omega S^{2np^j+1}.$$

Thus the diagram (9.1.2) can be enlarged to

$$
\begin{array}{ccc}
\Omega S^{2np^j+1} & \xrightarrow{\;\cong\;} & \Omega S^{2np^j+1} \\
\uparrow & & \uparrow H \\
\Omega SU(n) \longrightarrow \Omega SU(n+1) & \longrightarrow & \Omega S^{2n+1} \\
\cong\uparrow & \uparrow & \uparrow \\
\Omega SU(n) \longrightarrow B_{p^j-1} & \longrightarrow & J_{p^j-1}S^{2n} \\
& \uparrow & \uparrow \\
\Omega^2 S^{2np^j+1} & \xrightarrow{\;\cong\;} & \Omega^2 S^{2np^j+1}
\end{array}
$$

in which each row and column is a fibre sequence.

Of particular interest is the map

$$\Omega^2 S^{2np^j+1} \longrightarrow B_{p^j-1}.$$

We can think of the double loop space $\Omega^2 S^{2np^j+1}$ as a topological group acting on the space B_{p^j-1}, so there is an action map

$$\Omega^2 S^{2np^j+1} \times B_{p^j-1} \longrightarrow B_{p^j-1}. \tag{9.5.6}$$

Recall that G_j is the Thom spectrum of a certain stable vector bundle over B_{p^j-1}. This means that (9.5.6) leads to a stable map

$$\Sigma^\infty \Omega^2 S_+^{2np^j+1} \wedge G_j \xrightarrow{\mu} G_j. \tag{9.5.7}$$

Here we are skipping over some technical details which can be found in [DHS88, §3].

The space $\Omega^2 S^{2np^j+1}$ was shown by Snaith [Sna74] to have a stable splitting, which was described in 9.4.8. After localizing at p, this splitting has the form

$$\Sigma^\infty \Omega^2 S_+^{2np^j+1} \simeq (S^0 \vee S^{2np^j-1}) \wedge \bigvee_{i \geq 0} \Sigma^{i|b_{n,j}|} D_i$$

where each D_i is a certain finite complex (independent of n and j) with bottom cell in dimension 0. Moreover there are maps

$$S^0 = D_0 \xrightarrow{\ell} D_1 \xrightarrow{\ell} D_2 \xrightarrow{\ell} \cdots$$

of degree 1 on the bottom cell, and the limit, $\lim_{\rightarrow} D_i$, is known (9.4.10) to be the mod p Eilenberg-Mac Lane spectrum $H/(p)$.

In [DHS88, Prop. 3.19] it is shown that our map $b_{n,j}$ is the composite

$$\Sigma^{|b_{n,j}|} G_j \longrightarrow \Sigma^{|b_{n,j}|} D_1 \wedge G_j \longrightarrow \Sigma^{\infty} \Omega^2 S_+^{2np^j+1} \wedge G_j \xrightarrow{\mu} G_j.$$

and $b_{n,j}^m$ is the composite

$$\Sigma^{m|b_{n,j}|} G_j \longrightarrow \Sigma^{m|b_{n,j}|} D_m \wedge G_j \longrightarrow \Sigma^{\infty} \Omega^2 S_+^{2np^j+1} \wedge G_j \xrightarrow{\mu} G_j.$$

Thus we get a diagram

$$
\begin{array}{ccccccc}
G_j & \xrightarrow{\iota \wedge G_j} & D_1 \wedge G_j & \xrightarrow{\iota \wedge G_j} & D_2 \wedge G_j & \longrightarrow & \cdots \\
\cong \downarrow & & \mu \downarrow & & \mu \downarrow & & \\
G_j & \xrightarrow{b_{n,j}} & \Sigma^{-|b_{n,j}|} G_j & \xrightarrow{b_{n,j}} & \Sigma^{-2|b_{n,j}|} G_j & \longrightarrow & \cdots
\end{array}
$$

$$(9.5.8)$$

This means that the map

$$G_j \longrightarrow b_{n,j}^{-1} G_j$$

factors through $G_j \wedge H/(p)$.

Now consider the diagram

$$
\begin{array}{ccccc}
G_j & \longrightarrow & G_j \wedge H/(p) & \longrightarrow & b_{n,j}^{-1} G_j \\
b_{n,j} \downarrow & & b_{n,j} \wedge H/(p) \downarrow & & \downarrow \cong \\
\Sigma^{-|b_{n,j}|} G_j & \longrightarrow & \Sigma^{-|b_{n,j}|} G_j \wedge H/(p) & \longrightarrow & b_{n,j}^{-1} G_j \\
b_{n,j} \downarrow & & b_{n,j} \wedge H/(p) \downarrow & & \downarrow \cong \\
\vdots & & \vdots & & \vdots
\end{array}
$$

The middle vertical map is null because $b_{n,j}$ induces the trivial map in homology. Passing to the limit, we get

$$b_{n,j}^{-1} G_j \longrightarrow \text{pt.} \longrightarrow b_{n,j}^{-1} G_j$$

with the composite being the identity map on the telescope $b_{n,j}^{-1} G_j$. This shows that the telescope is contractible as desired. ∎

9.6   Historical note: theorems of Nishida and Toda

The method used to prove 9.5.5 is similar to ones used earlier by Toda and Nishida. Nishida's theorem (2.2.5, [Nis73]) was the special case of the nilpotence theorem (1.4.2) where X is the sphere spectrum. It was an important motivation for conjecturing the nilpotence theorem. Nishida's work was in turn inspired by the *extended power construction* introduced by Toda in [Tod68].

We will sketch part of Nishida's argument. Suppose $\alpha \in \pi_{2k}(S^0)$ has order p. We wish to show that it is nilpotent. (There is no loss of generality in assuming that the dimension of α is even since we could replace α by α^2.) The fact that α has order p means we have an extension

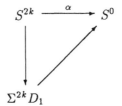

where D_1 is as in 9.4.8.

The extended power construction generalizes this to an extension

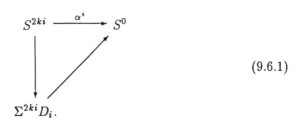

$$(9.6.1)$$

This is similar to the extension

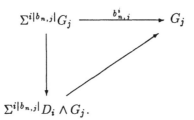

given by (9.5.8). We did not need to introduce the extended power construction there, because nature provided it for us in the form of the Snaith splitting 9.4.8.

Now we know from 9.4.10 that the map $D_i \to H/p$ is an equivalence through a range of dimensions that increases with i. (The computations used to prove 9.4.10 can be used to find this range precisely.) We can choose $i \gg 0$ so that this range exceeds $2k$. Now consider the diagram

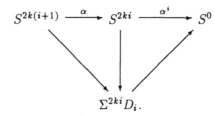

The map $S^{2k(i+1)} \to \Sigma^{2ki} D_i$ is null since the target has no homotopy in that dimension. It follows that $\alpha^{i+1} = 0$ and α is nilpotent, so we have proved the following special case of Nishida's theorem 2.2.5.

Theorem 9.6.2 *If $\alpha \in \pi_{2k}(S^0)$ has order p then it is nilpotent.*

Prior to Nishida's work, Toda [Tod68] used (9.6.1) in the case $i = p$. D_p is a 4-cell complex of the form

$$D_p = S^0 \cup_p e^1 \cup_{\alpha_1} e^{2p-2} \cup_p e^{2p-1}.$$

The notation is meant to suggest that the $(2p - 2)$-cell is attached to the 0-cell by

$$\alpha_1 \in \pi_{2p-3}(S^0)$$

and the $(2p - 1)$-cell is attached to the $(2p - 2)$-cell by a map of degree p.

It follows that the composite

$$S^{2p-3} \xrightarrow{\alpha_1} S^0 \longrightarrow D_p$$

is null. This means that for α as above, the composite

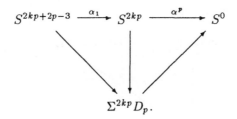

is null, i.e., we have proved

Theorem 9.6.3 (Toda's theorem) *If $\alpha \in \pi_{2k}(S^0)$ has order p for p odd then*

$$\alpha_1 \alpha^p = 0.$$

The first such α is

$$\beta_1 \in \pi_{2p^2-2p-2}(S^0)$$

so 9.6.3 gives

$$\alpha_1 \beta_1^p = 0, \tag{9.6.4}$$

which was also proved in [Tod67]. This relation does *not* hold in the E_2-term of either the classical Adams spectral sequence or the Adams-Novikov spectral sequence so (9.6.4) gives a nontrivial differential in each of them. In the case of the Adams-Novikov spectral sequence for an odd prime, it is the *first* nontrivial differential.

Appendix A

Some tools from homotopy theory

In this appendix we will give some more background on many of the results quoted in the text. Before we begin we list some standard references in the subject.

Eilenberg-Steenrod [ES52] is the first modern treatment of algebraic topology. They introduce axioms for homology and cohomology. Forty years after its publication it is still worth looking at. Spanier [Spa66] is useful as a reference. Less ambitious but more readable are the books by Vick [Vic73] and Greenberg-Harper [GH81]. Gray's book [Gra75] introduces the subject from a viewpoint more in tune with the way working homotopy theorists actually think about it. The texts by Dold [Dol80] and Switzer [Swi75] also cover all of the standard facts about ordinary homology and CW-complexes. The latter treats some more modern topics such as K-theory and the Adams spectral sequence. Whitehead's book [Whi78] is a thorough introduction to homotopy theory from an elementary point of view. For more advanced topics, Adams' books [Ada74] and [Ada78] are invaluable.

A.1 CW-complexes

Definition A.1.1 *A* **CW-complex** X *is a topological space built up out of subspaces*

$$X^0 \subset X^1 \subset X^2 \subset \cdots \subset X$$

called **skeleta** *in the following way.* X^0 *is a discrete set of points, possibly infinite.* X^k *is obtained from* X^{k-1} *by the process of* **adjoining k-cells**.

119

One has a (possibly infinite) collection of k-dimensional balls $\{B_\alpha^k\}$ (called k-cells*) bounded by spheres $\{S_\alpha^{k-1}\}$ and* **attaching maps**

$$f_\alpha : S_\alpha^{k-1} \longrightarrow X^{k-1};$$

X^k *is the quotient of the disjoint union of X^{k-1} with the balls B_α^k obtained by identifying the boundaries of these balls with their images in X_{k-1} under the attaching maps. Equivalently, X^k is the cofibre of the map*

$$\coprod S_\alpha^{k-1} \xrightarrow{\ \coprod f_\alpha\ } X^{k-1}.$$

A **bottom cell** *of X is one whose dimension is minimal among all positive dimensional cells of X. (If this dimension is k and X is path-connected, then X is $(k-1)$-**connected**, i.e., all homotopy groups below dimension k vanish.) X has* **finite type** *if the number of cells in each dimension, including zero (i.e., the number of points in X^0), is finite. X is* **finite** *if the total number of cells is finite. In this case a* **top cell** *of X is one whose dimension is maximal. Its dimension will be referred to as the* **dimension of** X.

Example A.1.2 (The torus as a CW-complex) *The 2-dimensional torus $T^2 = S^1 \times S^1$ is a CW-complex as follows. X^0 is a single point and there are two 1-cells, so $X^1 = S^1 \vee S^1$. Recall that the torus can be regarded as the quotient obtained from the unit square by identifying opposite pairs of edges in an orientation preserving way. This identification can be thought of as a map f from the boundary of the square (i.e., S^1), to X^1. This f is the attaching map for a single 2-cell, and its cofibre is the torus T^2.*

A simplicial complex is a special case of a CW-complex in which the attaching maps are required to have certain special properties. Hence objects which can be triangulated, such as smooth or piecewise linear manifolds, algebraic varieties over the real or complex numbers with the standard (rather than the Zariski) topology, and classifying spaces of Lie groups, are CW-complexes. On the other hand the Cantor set, the rational numbers, and algebraic varieties with the Zariski topology, are not CW-complexes, not even up to homotopy equivalence. The class of spaces having the homotopy type of a CW-complex is large enough to include all the spaces one ever considers in homotopy theory.

The following two results indicate how convenient CW-complexes are. The first is proved in [Spa66, 7.6.24].

Theorem A.1.3 *Let $f : X \to Y$ be a continuous map between path-connected CW-complexes. If f induces an isomorphism in π_i for each $i > 0$, then f is a homotopy equivalence.*

The next result is proved by Milnor in [Mil59].

Theorem A.1.4 *Let X and Y (with X compact) be spaces each homotopy equivalent to a CW-complex . Then the same is true of the function space*

$$\text{Map}(X, Y),$$

the space of continuous maps from X to Y with the compact-open topology.

This is surprising since such function spaces tend to be infinite dimensional when X and Y are finite CW-complexes. A similar result holds for the space of maps sending a prescribed collection of subspaces (each homotopy equivalent to a CW-complex) of X to a similar collection of subspaces of Y.

An important example of this phenomenon, proved earlier by James [Jam55], concerns the loop space ΩS^{n+1} (defined below in A.2.1) i.e., the space of base point preserving maps from the circle S^1 to S^{n+1}.

Theorem A.1.5 *The loop space ΩS^{n+1} (for $n > 0$) is homotopy equivalent to a CW-complex of the form*

$$S^n \cup e^{2n} \cup e^{3n} \cup \cdots,$$

i.e., one with a single cell in each dimension divisible by n.

Now we will state a result similar in spirit to 2.3.4(i).

Proposition A.1.6 *Let*

$$X \xrightarrow{f} Y \xrightarrow{i} C_f \xrightarrow{j} \Sigma X \xrightarrow{\Sigma f} \Sigma Y \longrightarrow \cdots$$

be a cofibre sequence (as in 2.3.3) in which X and Y are each $(k-1)$-connected (i.e., neither has any positive dimensional cells below dimension k) and let W be a finite CW-complex which is a double suspension with top cell in dimension less than $2k - 1$. Then there is a long exact sequence of abelian groups

$$[W, X] \xrightarrow{f_*} [W, Y] \xrightarrow{i_*} [W, C_f] \xrightarrow{j_*} [W, \Sigma X] \xrightarrow{\Sigma f_*} [W, \Sigma Y] \longrightarrow \cdots.$$

This sequence will terminate at the point where the connectivity of the target exceeds the dimension of W.

A.2 Loop spaces and spectra

Now we turn to spectra, which were defined in 5.1.1. In order to give a better definition of a map of spectra, we must first describe loop spaces and adjoint maps.

Definition A.2.1 *The* **loop space** *of* X, ΩX *is the space of basepoint-preserving maps of the circle* S^1 *into* X, *with the compact-open topology. The* i^{th} **iterated loop space** *of* X, $\Omega^i X$ *is defined inductively by* $\Omega(\Omega^{i-1}X)$. *Equivalently, it is the space of base point preserving maps of* S^i *into* X, *with the compact-open topology.*

Definition A.2.2 *Given a map* $f: \Sigma X \to Y$, *for each point in* X *we get a closed path in* Y, *since the suspension* ΣX *is a quotient of* $X \times [0,1]$ *(1.3.1). The resulting map* $\hat{f}: X \to \Omega Y$ *is the* **adjoint** *of* f. *Similarly, a map* $\Sigma^i X \to Y$ *is adjoint to a map* $X \to \Omega^i Y$.

Proposition A.2.3 *The construction above gives a one-to-one correspondence between maps* $\Sigma^i X \to Y$ *and maps* $X \to \Omega^i Y$ *and an isomorphism*

$$[\Sigma^i X, Y] \longrightarrow [X, \Omega^i Y].$$

In particular we have

$$\pi_{k+i}(Y) \cong \pi_k(\Omega^i Y).$$

Now recall (5.1.1) that a spectrum E is defined to be a collection of spaces $\{E_n\}$ and maps $\Sigma E_n \to E_{n+1}$; we say that E is a **suspension spectrum** if each of these maps is an equivalence. Each such map is adjoint to a map $E_n \to \Omega E_{n+1}$; if these are all equivalences, we say that E is an Ω-**spectrum**. In this case it follows that E_n is homotopy equivalent to $\Omega^k E_{n+k}$ for each $k > 0$. In particular, each E_n is an **infinite loop space**. Such spaces play a special role in homotopy theory; for more information, see Adams' excellent account, [Ada78].

The most familiar example of an Ω-spectrum is the Eilenberg-Mac Lane spectrum, HA (where A is an abelian group) defined by $(HA)_n = K(A, n)$.

If E and F are spectra with F an Ω-spectrum, then a map $f: E \to F$ is precisely what one would expect: a collection of maps $f_n: E_n \to F_n$ such that $f_n = \Omega f_{n+1}$.

However, for more general spectra, this definition is far too restrictive. Here is a simple example which illustrates this point. Let E be the Eilenberg-Mac Lane spectrum HA and let the spectrum F be defined by

$$F_n = K(A, n)^{2n},$$

the $(2n)$-skeleton of $K(A, n)$. Then there is a map $i : F \to E$ induced by the evident inclusions. It is easy to show that it induces an isomorphism of homotopy groups, and should therefore be a homotopy equivalence. This means that there should be some sort of inverse map from E to F. On the other hand, there is no map $E_n \to F_n$ with suitable properties, i.e. which is an equivalence below dimension $2n$.

One way out of this difficulty is to replace the target spectrum F by a homotopy equivalent Ω-spectrum $\tilde{F}$ as follows: set $\tilde{F}_n$ equal to

$$\lim_{\to} \Omega^k F_{n+k}. \tag{A.2.4}$$

Then the evident maps $F_n \to \tilde{F}_n$ can be shown to map the homotopy groups of F isomorphically to those of $\tilde{F}$, so the two spectra are homotopy equivalent. Moreover we have isomorphisms

$$\pi_k(F) \cong \pi_{n+k}(\tilde{F}_n)$$

whenever $n + k > 0$. On the other hand, $\overline{H}(\tilde{F}_n)$ bears little resemblance to $H_*(F)$ for any n, or to $\overline{H}_*(F_n)$ if F is not an Ω-spectrum.

Hence we can make the following definition.

Definition A.2.5 *A map of spectra* $f \colon E \to F$ *is a collection of maps*

$$E_n \xrightarrow{f_n} \lim_{\substack{\to \\ k}} \Omega^k F_{n+k}$$

with $f_n = \Omega f_{n+1}$.

We remark that in the case where E and F are suspension spectra, then each such collection of maps is equivalent to one induced from a single map $E_n \to F_n$ for some n.

Following Adams [Ada74, III.2], we will work in the follwing category.

Definition A.2.6 *The* **homotopy category of CW-spectra** *is the category whose objects are CW-spectra (i.e. spectra as defined in 5.1.1 in which all spaces in sight have the homotpy type of CW-complexes), and whose morphisms are homotopy classes of maps, as defined above.*

One can form *coproducts* in the category of spectra as follows. Recall that the coproduct in the category of pointed spaces is the one point union or wedge. This means that given any collection $\{X_\alpha\}$ of pointed spaces with pointed maps $f_\alpha : X_\alpha \to Y$, we get a unique map

$$\bigvee_\alpha X_\alpha \xrightarrow{f} Y.$$

If we have spectra X_α we can define their coproduct by

$$(\bigvee_\alpha X_\alpha)_n = \bigvee_\alpha (X_\alpha)_n, \qquad\qquad (A.2.7)$$

and in view of our definition of a map of spectra (A.2.5), a collection of maps $f_\alpha : X_\alpha \to Y$ leads to a unique map $f : \bigvee_\alpha X_\alpha \to Y$. Moreover we have

$$E_*(\bigvee_\alpha X_\alpha) = \bigoplus_\alpha E_*(X_\alpha).$$

We can also define products in the category of spectra; this will be done below in A.4.3.

Definition A.2.8 *A* **ring spectrum** *E is a spectrum equipped with maps $\eta: S^0 \to E$, called the* **unit map**, *and $m: E \wedge E \to E$, called the* **multiplication map**, *such that the composites*

$$E = S^0 \wedge E \xrightarrow{\quad \eta \wedge E \quad} E \wedge E \xrightarrow{\quad m \quad} E \quad and$$
$$E = E \wedge S^0 \xrightarrow{\quad E \wedge \eta \quad} E \wedge E \xrightarrow{\quad m \quad} E$$

are each the identity on E (this is analogous to the unitary condition on a ring), and the following diagram commutes up to homotopy.

$$
\begin{array}{ccc}
E \wedge E \wedge E & \xrightarrow{\;m \wedge E\;} & E \wedge E \\
{\scriptstyle E \wedge m}\big\downarrow & & \big\downarrow{\scriptstyle m} \\
E \wedge E & \xrightarrow{\quad m \quad} & E
\end{array}
$$

This is an associativity condition on m. If the multiplication m is commutative up to homotopy, then E is **homotopy commutative**.

A **module spectrum** *M over E is one equipped with a map*

$$E \wedge M \xrightarrow{\;\mu\;} M$$

such that the following diagram commutes up to homotopy.

$$
\begin{array}{ccc}
E \wedge E \wedge M & \xrightarrow{\;m \wedge E\;} & E \wedge M \\
{\scriptstyle E \wedge \mu}\big\downarrow & & \big\downarrow{\scriptstyle \mu} \\
E \wedge M & \xrightarrow{\quad \mu \quad} & M
\end{array}
$$

Definition A.2.9 *A ring spectrum E is* **flat** *if $E \wedge E$ is equivalent to a wedge of suspensions of E.*

Under these circumstances, $\pi_*(E)$ is a ring and $\pi_*(M)$ is a module over it.

An element $v \in \pi_d(E)$ is represented by a map $f \colon S^d \to E$. Using the multiplication on E we have the composite

$$\Sigma^d E = S^d \wedge E \xrightarrow{\quad f \wedge E \quad} E \wedge E \xrightarrow{\ m\ } E$$

and this composite induces multiplication by v in homotopy. We will denote this map also by f. The following definition involves a direct limit of spectra. These will be discussed below in A.5.

Definition A.2.10 *With notation as above, $v^{-1}E$ is the direct limit or telescope of*

$$E \xrightarrow{\ f\ } \Sigma^{-d} E \xrightarrow{\ f\ } \Sigma^{-2d} E \xrightarrow{\ f\ } \cdots .$$

(Its homotopy is

$$v^{-1} E_* = E_* \otimes_{\mathbf{Z}[v]} \mathbf{Z}[v, v^{-1}]$$

and it is a module spectrum over E.)

Recall that a spectrum is connective if its homotopy groups vanish below some dimension. Every suspension spectrum is connective. The inexperienced reader may find that some of his intuition about spaces fails him when dealing with nonconnective spectra. We will now give an important example which partly illustrates this point.

Let X be a p-local finite complex of type n (with $n > 0$) (1.5.3) with a v_n-map (1.5.4)

$$\Sigma^d X \xrightarrow{\ f\ } X.$$

We will denote the corresponding suspension spectrum by X also. Since spectra can be desuspended any number of times, we can form a directed system

$$X \xrightarrow{\ f\ } \Sigma^{-d} X \xrightarrow{\ f\ } \Sigma^{-2d} X \xrightarrow{\ f\ } \cdots \qquad (A.2.11)$$

We want to look at the homotopy direct limit of this system. Such limits are described below in A.5. We define

$$\hat{X} = \varinjlim \Sigma^{-di} X, \qquad (A.2.12)$$

which we call the *telescope of f*.

Proposition A.2.13 *If $\hat{X}$ is the telescope defined above, then*

(i) $K(n)_*(\hat{X}) = K(n)_*(X)$,

(ii) $K(m)_*(\hat{X}) = 0$ *for* $m \neq n$ *and*

(iii) $H_*(\hat{X}) = 0$.

Proof. Generalized homology commutes with homotopy direct limits (A.5.7). By assumption, $K(n)_*(f)$ is an isomorphism and $K(m)_*(f) = 0$ for $m \neq n$, which proves (i) and (ii). From 1.5.2(vi) we see that $H_*(f) = 0$, which proves (iii). ∎

Hence $\hat{X}$, which is nonconnective, has trivial ordinary homology but it is not contractible, since it has nontrivial Morava K-theory. On the other hand, any simply connected CW-complex with trivial homology is contractible, and the same is true of any connective spectrum. There is a Hurewicz theorem indexHurewicz theorem for connective spectra which says that the first nontrivial homology and homotopy groups are isomorphic. There is no such theorem for nonconnective spectra, where there is no first nontrivial homotopy group.

A.3   Generalized homology and cohomology theories

Now we will discuss generalized homology theories. First we need to recall some facts about ordinary homology, which is described in detail in any textbook on algebraic topology.

For each space X one has a graded abelian group $H_*(X)$, i.e., an abelian group $H_i(X)$ for each integer i. (These groups vanish for negative i, but in generalized homology this need not be the case.) Given a nonempty subspace $A \subset X$ one has **relative homology groups** $H_*(X, A)$. In positive dimensions these are, under mild hypotheses, the same as $H_*(X/A)$, where X/A denotes the topological quotient of X obtained by shrinking A to a single point. A map $f: (X, A) \to (Y, B)$ is a continuous map from X to Y that sends $A \subset X$ to $B \subset Y$. It induces a homomorphism $f_*: H_*(X, A) \to H_*(Y, B)$.

A classical theorem of algebraic topology due to Eilenberg and Steenrod [ES52] says that ordinary homology theory is characterized by the following axioms.

A.3.1 Eilenberg-Steenrod axioms
(a) Homotopy axiom: *Homotopic maps $f, g: (X, A) \to (Y, B)$ induce the same homomorphism $H_*(X, A) \to H_*(Y, B)$.*

(b) Exactness axiom: *For each pair (X, A) there is a natural long exact sequence*

$$\cdots \longrightarrow H_n(A) \xrightarrow{i_*} H_n(X) \xrightarrow{j_*} H_n(X, A) \xrightarrow{\partial} H_{n-1}(A) \longrightarrow \cdots$$

where i_* is the homomorphism induced by the inclusion map $i: A \to X$. Naturality means that given a map $f : (X, A) \to (Y, B)$, the following diagram commutes.

$$
\begin{array}{ccccccc}
H_n(A) & \xrightarrow{i_*} & H_n(X) & \xrightarrow{j_*} & H_n(X, A) & \xrightarrow{\partial} & H_{n-1}(A) \\
\downarrow{\scriptstyle f_*} & & \downarrow{\scriptstyle f_*} & & \downarrow{\scriptstyle f_*} & & \downarrow{\scriptstyle f_*} \\
H_n(B) & \xrightarrow{i_*} & H_n(Y) & \xrightarrow{j_*} & H_n(Y, B) & \xrightarrow{\partial} & H_{n-1}(B).
\end{array}
$$

(c) Excision axiom: *If $C \subset A \subset X$ with the closure of C contained in the interior of A, there is an isomorphism*

$$H_*(X - C, A - C) \xrightarrow{\cong} H_*(X, A).$$

(d) Dimension axiom: *When X is a single point we have*

$$H_i(X) = \begin{cases} \mathbf{Z} & \text{if } i = 0 \\ 0 & \text{otherwise.} \end{cases}$$

axioms for $H_*(X; G)$, the homology of X with coefficients in an abelian group G, can be obtained by modifying the Dimension axiom. There are similar axioms for cohomology, obtained from the above by reversing all the arrows.

If G is a ring R we have **cup products** in $H^*(X; R)$, i.e., given $u \in H^i(X)$ and $v \in H^j(X)$, their cup product $u \cup v$ (usually denoted simply by uv) is defined in $H^{i+j}(X; R)$. This product is commutative up to sign, i.e.,

$$vu = (-1)^{ij} uv.$$

It comes from the composite

$$H^*(X) \otimes H^*(X) \xrightarrow{\kappa} H^*(X \times X) \xrightarrow{\Delta^*} H^*(X)$$

where Δ^* is induced by the diagonal embedding $\Delta X \to X \times X$ and κ is the Künneth homomorphism. The latter is an isomorphism if $H^*(X)$ is flat as an R-module (in particular, if R is a field or if $H^*(X; \mathbf{Z})$ is torsion free) but not in general. One also has cup product pairings in relative cohomology, namely

$$H^*(X) \otimes H^*(X, A) \longrightarrow H^*(X, A) \qquad \text{(A.3.2)}$$

and

$$H^*(X, A) \otimes H^*(X, B) \longrightarrow H^*(X, A \cup B)$$

The **reduced homology** of a space $\overline{H}_*(X)$ is the kernel of the map $H_*(X) \to H_*(\text{pt.})$ and the **reduced cohomology** $\overline{H}^*(X)$ is the cokernel of the map $H^*(X) \leftarrow H^*(\text{pt.})$.

The following definition is due to G. W. Whitehead [Whi62]

Definition A.3.3 *A* **generalized homology theory** h_* *is a covariant functor from the category of CW-complexes (or pairs thereof) to the category of graded abelian groups that satisfies the first three of the Eilenberg-Steenrod axioms. A* **generalized cohomology theory** h^* *is a contravariant functor with similar properties.*

Such theories can be constructed in the following way.

Definition A.3.4 *Let E be an Ω-spectrum. The* **generalized cohomology theory associated with** E, E^*, *is defined by*

$$E^n(X) = [X, E_n]$$

and the **generalized homology theory associated with** E, E_*, *is defined by*

$$E_n(X) = \pi_n(E \wedge X)$$

where $E \wedge X$ denotes the smash product (5.1.3) of E with the suspension spectrum associated with X.

Such a theory is **multiplicative** *if E is a ring spectrum (A.2.8). In that case there is a* **Hurewicz map**

$$\pi_*(X) = \pi_*(S^0 \wedge X) \xrightarrow{h} E_*(X) = \pi_*(E \wedge X)$$

induced by the unit map $\eta: S^0 \to E$.

Notice that if X is also a ring spectrum then $\pi_*(X)$ and $E_*(X)$ have natural ring structures and h is a ring homomorphism.

In particular ordinary homology and cohomology can be defined in this way by taking E to be the Eilenberg-MacLane spectrum H. If E is the sphere spectrum S, the resulting homology theory is stable homotopy,

$\pi_*^S(X)$. Another well known example is classical complex K-theory; the Ω-spectrum K is defined by

$$K_n = \left\{ \begin{array}{ll} \mathbf{Z} \times BU & \text{if } n \text{ is even} \\ U & \text{if } n \text{ is odd,} \end{array} \right.$$

where U is the stable unitary group and BU is its classifying space.

A generalized cohomology theory E^* has cup products similar to the ones in ordinary cohomology provided that E is a ring spectrum (A.2.8). The multiplication on E, and therefore the product in $E^*(X)$, need not be commutative, even up to sign. For example, the Morava K-theories at the prime 2 are noncommutative.

Theories constructed using A.3.4 also satisfy the following axiom.

A.3.5 (Wedge axiom) *If W is a (possibly infinite) wedge of spaces $\vee X_\alpha$ then*

$$h^*(W) \cong \prod h^*(X_\alpha)$$

and

$$h_*(W) \cong \bigoplus h_*(X_\alpha).$$

Note that for finite wedges this statement is a consequence of the Eilenberg-Steenrod axioms.

The best tool for computing $h_*(X)$ and $h^*(X)$ for a space or connective spectrum X is the *Atiyah-Hirzebruch spectral sequence*. A more detailed account is given by Adams in [Ada74, III.7].

For homology it is constructed as follows. Assume for simplicity that X is (-1)-connected if it is a spectrum. It has a skeletal filtration

$$X^0 \subset X^1 \subset X^2 \subset \cdots$$

and each subquotient X^n/X^{n-1} is a wedge of n-spheres. It follows that

$$h_*(X^n/X^{n-1}) = h_*(\text{pt.}) \otimes H_*(X^n/X^{n-1}). \qquad (\text{A.3.6})$$

Thus for each n there is a long exact sequence

$$\longrightarrow h_*(X^{n-1}) \longrightarrow h_*(X^n) \longrightarrow h_*(X^n/X^{n-1}) \longrightarrow$$

in which every third term is known. These can be assembled into an exact couple which gives the desired spectral sequence. Its E_1-term consists of the groups given in (A.3.6), which depends on the choice of skeleta. However its E_2-term depends only on $H_*(X)$.

Theorem A.3.7 *Let X be a space or a connective spectrum.*

(a) For any generalized homology theory h_* there is a spectral sequence converging to $h_*(X)$ with

$$E_2^{s,t} = H_s(X; h_t(\mathrm{pt.}))$$

and

$$E_r^{s,t} \xrightarrow{d_r} E_r^{s-r,t+r-1}.$$

If h_* is multiplicative and X is a ring spectrum or an H-space, then this is a spectral sequence of algebras, i.e., each d_r is a derivation.

(b) For any generalized cohomology theory h^* there is a spectral sequence converging to $h^*(X)$ with

$$E_2^{s,t} = H^s(X; h^t(\mathrm{pt.}))$$

and

$$E_r^{s,t} \xrightarrow{d_r} E_r^{s+r,t-r+1}.$$

If h^* is multiplicative and X is a space, then this is a spectral sequence of algebras, i.e., each d_r is a derivation.

A.4 Brown representability

The following extremely useful result is due to E. H. Brown [Bro62]. A simplified proof is given by Spanier in [Spa66]. The theorem was strengthened by Adams in [Ada71].

Theorem A.4.1 (Brown representability theorem) *If h^* is a generalized cohomology theory satisfying the first three Eilenberg-Steenrod axioms A.3.1 and the wedge axiom A.3.5 then there is a spectrum E such that $h^* = E^*$, and similarly for generalized homology theories.*

Adams' generalization requires h^* to be defined only on the category of *finite* CW-complexes.

This theorem has been used to construct spectra by constructing the cohomology theory it represents. For example, given spectra X and Y, the graded group

$$[W \wedge X, Y]_*,$$

regarded as a functor on the spectrum W, is a cohomology theory satisfying the wedge axiom. Therefore by A.4.1 there is a spectrum, denoted by $F(X, Y)$ and called the **function spectrum**, such that

$$[W \wedge X, Y]_* \cong [W, F(X, Y)]_*. \tag{A.4.2}$$

When X is finite and $Y = S^0$, then $F(X, Y)$ is the Spanier-Whitehead dual of X, DX.

We can also use A.4.1 to define products of spectra.

Proposition A.4.3 *Given any collection of spectra $\{X_\alpha\}$ there is a product spectrum $\prod_\alpha X_\alpha$ satisfying*

$$[Y, \prod_\alpha X_\alpha]_* = \prod_\alpha [Y, X_\alpha]_*$$

for any spectrum Y.

If there are only a finite number of factors, then the product is the same as the coproduct, i.e., the wedge defined in (A.2.7).

Proof of A.4.3.   The expression on the right, viewed as a functor of Y, satisfies Brown's axioms (i.e., the ones of A.4.1). It therefore has a representing spectrum, which is the desired product.

Alternatively, assuming that each X_α in an Ω-spectrum, we could define $\prod_\alpha X_\alpha$ explicilty by setting

$$(\prod_\alpha X_\alpha)_n = \prod_\alpha (X_\alpha)_n$$

and use A.2.5 (and the fact that looping commutes with infinite Cartesian products) to show that it has the desired property.   ■

A.5   Limits in the stable homotopy category

In this section we will give a brief review of homotopy direct and inverse limits. A more detailed account can be found in Bousfield-Kan [BK72].

Direct limits of abelian groups

First recall the definition of the direct limit of abelian groups. Suppose we have groups and homomorphisms

$$A_1 \xrightarrow{f_1} A_2 \xrightarrow{f_2} A_3 \xrightarrow{f_3} \cdots$$

We have the *shift homomorphism*

$$\bigoplus_{i>0} A_i \xrightarrow{s} \bigoplus_{i>0} A_i \tag{A.5.1}$$

defined by

$$s(a_i) = a_i - f_i(a_i)$$

for $a_i \in A_i$. It is always a monomorphism. The direct limit is defined by

$$\lim_{\to} A_i = \operatorname{coker} s. \qquad (A.5.2)$$

It has a universal property; for any collection of homomorphisms $g_i : A_i \to B$ compatible under the f_i there is a unique homomorphism $g : \lim_{\to} A_i \to B$ such that for each i the composite

$$A_i \longrightarrow \lim_{\to} A_i \overset{g}{\longrightarrow} B \qquad (A.5.3)$$

is g_i. To see this, note that the compatibility condition amounts to requiring that the composite

$$\bigoplus_{i>0} A_i \overset{s}{\longrightarrow} \bigoplus_{i>0} A_i \overset{\oplus g_i}{\longrightarrow} B$$

be trivial, so $\oplus g_i$ factors uniquely through coker s.

Proposition A.5.4 *Direct limits are exact, i.e., if we have a commutative diagram*

$$
\begin{array}{ccccccc}
0 & & 0 & & 0 & & \\
\downarrow & & \downarrow & & \downarrow & & \\
A_1 & \longrightarrow & A_2 & \longrightarrow & A_3 & \longrightarrow & \cdots \\
\downarrow & & \downarrow & & \downarrow & & \\
B_1 & \longrightarrow & B_2 & \longrightarrow & B_3 & \longrightarrow & \cdots \\
\downarrow & & \downarrow & & \downarrow & & \\
C_1 & \longrightarrow & C_2 & \longrightarrow & C_3 & \longrightarrow & \cdots \\
\downarrow & & \downarrow & & \downarrow & & \\
0 & & 0 & & 0 & &
\end{array}
$$

where each column is exact, then we get a short exact sequence

$$0 \longrightarrow \lim_{\to} A_i \longrightarrow \lim_{\to} B_i \longrightarrow \lim_{\to} C_i \longrightarrow 0.$$

Proposition A.5.5 *Direct limits commute with tensor products, i.e.,*

$$\lim_{\to}(A_i \otimes B) = (\lim_{\to} A_i) \otimes B.$$

Homotopy direct limits of spectra

Now we want to mimic this construction in homotopy theory. Let

$$X_1 \xrightarrow{f_1} X_2 \xrightarrow{f_2} X_3 \xrightarrow{f_3} \cdots$$

be a collection of spectra and continuous maps. As in (A.2.7) we can define the infinite coproduct or wedge of these spectra,

$$\bigvee_{i>0} X_i$$

with

$$\pi_*(\bigvee_{i>0} X_i) = \bigoplus_{i>0} \pi_*(X_i).$$

It distributes over smash products in the expected way, i.e.,

$$E \wedge (\bigvee_{i>0} X_i) = \bigvee_{i>0} (E \wedge X_i),$$

so

$$E_*(\bigvee_{i>0} X_i) = \bigoplus_{i>0} E_*(X_i).$$

Moreover, there is a *shift map*

$$\bigvee_{i>0} X_i \xrightarrow{\sigma} \bigvee_{i>0} X_i$$

inducing the shift homomorphism of (A.5.1) in homology.
Hence we can mimic (A.5.2) and define

$$\varinjlim X_i = C_\sigma, \tag{A.5.6}$$

the cofibre of σ. This gives

$$E_*(\varinjlim X_i) = \varinjlim E_*(X_i), \tag{A.5.7}$$

i.e., homology commutes with direct limits.

However, this limit does *not* have the universal property analogous to (A.5.3), i.e., compatible maps $g_i : X_i \to Y$ do not lead to a unique map $g : \varinjlim X_i \to Y$. Instead we have a long exact sequence

$$\cdots \longleftarrow \prod_{i>0} [X_i, Y]_* \xleftarrow{\sigma^*} \prod_{i>0} [X_i, Y]_* \xleftarrow{j^*} [\varinjlim X_i, Y]_* \longleftarrow \cdots$$

(We are using the fact that the group of maps from an infinite coproduct is the infinite product of the groups of maps, i.e., cohomology converts coproducts to products. This is essentially Brown's wedge axiom A.3.5.) The maps g_i give us an element in $\ker \sigma^*$ and therefore in im j^*. However, σ^* need not be onto, so j^* need not be one-to-one and we do not have a unique map $g : \lim_{\rightarrow} X_i \rightarrow Y$.

This means that $\lim_{\rightarrow} X_i$ is not a *categorical* direct limit or colimit. For this reason many authors, including Bousfield-Kan, use the notation hocolim instead of $\lim_{\rightarrow}$.

This construction can readily be generalized to other directed systems of spectra. One of these is particularly useful. Given a spectrum X, consider the set of all maps $f : F \rightarrow X$ with F finite (5.1.1). We will call such a map a *finite subspectrum of X*. These can be thought of as objects in a category in which a morphism $(F_1, f_1) \rightarrow (F_2, f_2)$ is a map $g : F_1 \rightarrow F_2$ with $f_1 = f_2 g$. This category is directed because any pair of maps $f_i : F_i \rightarrow X$ ($i = 1, 2$) can be factored through the evident map $F_1 \vee F_2 \rightarrow X$. It follows that there is a canonical map

$$\lim_{\rightarrow} F_\alpha \xrightarrow{\lambda} X.$$

Proposition A.5.8 *The map λ above is a weak homotopy equivalence for any spectrum X, i.e., every CW-spectrum is the homotopy direct limit of its finite subspectra.*

Sketch of proof. We need to show that $\pi_*(\lambda)$ is an isomorphism. It is onto because every element of $\pi_*(X)$ is induced by a map from a sphere (which is a finite spectrum) to X. To show that it is one-to-one, let $x \in \ker \lambda$. It is represented by as a map g from a sphere to some finite subspectrum F of X, i.e., we have

$$S^n \xrightarrow{g} F \xrightarrow{f} X$$

with fg null. It follows that f factors through the mapping cone C_g. Then x has trivial image in $\pi_*(C_g)$ and hence in $\pi_*(\lim_{\rightarrow} F_\alpha)$.   ∎

Inverse limits of abelian groups

Homotopy inverse limits are defined in a similar way, once we know that there are infinite products in the homotopy category of spectra. Again we begin by recalling the definition for abelian groups. Given

$$A_1 \xleftarrow{f_1} A_2 \xleftarrow{f_2} A_3 \xleftarrow{f_3} \cdots,$$

there is a shift homomorphism

$$\prod_{i>0} A_i \xleftarrow{\ s\ } \prod_{i>0} A_i$$

defined by

$$s(a_1, a_2, a_3, \ldots) = (a_1 - f_1(a_2), a_2 - f_2(a_3), \ldots).$$

This map is neither one-to-one nor onto in general, and the inverse limit $\lim_{\leftarrow} A_i$ is $\ker s$ by definition. Its cokernel is denoted by $\lim_{\leftarrow}^1 A_i$. Thus we have a 4-term exact sequence

$$0 \longrightarrow \varprojlim A_i \longrightarrow \prod_{i>0} A_i \xrightarrow{\ s\ } \prod_{i>0} A_i \longrightarrow \varprojlim{}^1 A_i \longrightarrow 0.$$

Inverse limits have a universal property similar to that of direct limits, namely a collection of homomorphisms $g_i : B \to A_i$ with $f_i g_{i+1} = g_i$ induces a unique homomorphism $g : B \to \lim_{\to} A_i$.

The failure of the shift map s to be onto means that inverse limits do not preserve exactness. Instead we have the following result.

Proposition A.5.9 *If we have a commutative diagram*

$$
\begin{array}{ccccccc}
0 & & 0 & & 0 & & \\
\downarrow & & \downarrow & & \downarrow & & \\
A_1 & \longleftarrow & A_2 & \longleftarrow & A_3 & \longleftarrow & \cdots \\
\downarrow & & \downarrow & & \downarrow & & \\
B_1 & \longleftarrow & B_2 & \longleftarrow & B_3 & \longleftarrow & \cdots \\
\downarrow & & \downarrow & & \downarrow & & \\
C_1 & \longleftarrow & C_2 & \longleftarrow & C_3 & \longleftarrow & \cdots \\
\downarrow & & \downarrow & & \downarrow & & \\
0 & & 0 & & 0 & &
\end{array}
$$

where each column is exact, then we get a 6-term exact sequence

$$0 \longrightarrow \varprojlim A_i \longrightarrow \varprojlim B_i \longrightarrow \varprojlim C_i \longrightarrow \varprojlim{}^1 A_i \longrightarrow \varprojlim{}^1 B_i \longrightarrow \varprojlim{}^1 C_i \longrightarrow 0.$$

The group $\lim_{\leftarrow}^1$ is a nuisance. The following example is instructive.

Example A.5.10 *Consider the commutative diagram*

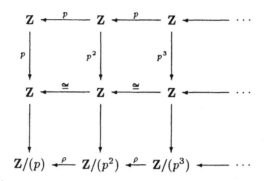

with exact columns. Then the inverse limit of the top row is trivial, while those of the lower two rows are the integers $\mathbf{Z}$ *and the p-adic integers* $\mathbf{Z}_p$. *It follows from A.5.9 that* $\lim^1_{\leftarrow}$ *for the first row is the group* $\mathbf{Z}_p/\mathbf{Z}$. *It is uncountable, and the topology it inherits from* $\mathbf{Z}_p$ *is the trivial one since* $\mathbf{Z}$ *is dense in* $\mathbf{Z}_p$.

If we tensor the bottom row with the rationals $\mathbf{Q}$, *we see that tensor products do not commute with inverse limits since*

$$\lim_{\leftarrow}(\mathbf{Z}/(p^i) \otimes \mathbf{Q}) = \lim_{\leftarrow}(0) = 0$$

but

$$\left(\lim_{\leftarrow} \mathbf{Z}/(p^i)\right) \otimes \mathbf{Q} = \mathbf{Z}_p \otimes \mathbf{Q} = \mathbf{Q}_p,$$

the p-adic numbers.

There are some simple conditions which guarantee that $\lim^1_{\leftarrow}$ vanishes.

Definition A.5.11 *An inverse system of abelian groups*

$$A_1 \longleftarrow A_2 \longleftarrow A_3 \longleftarrow \cdots$$

is **Mittag-Leffler** *if for each i the decreasing series of subgroups*

$$A_i \supset \operatorname{im} A_{i+j}$$

stabilizes after a finite number of steps, i.e., it is independent of j for $j \gg 0$.

Proposition A.5.12 *If the inverse system* $\{A_i\}$ *as above is Mittag-Leffler, then*

$$\lim_{\leftarrow}{}^1 A_i = 0.$$

In particular this is the case when

- *each group A_i is finite or*

- *each homomorphism $A_i \leftarrow A_{i+1}$ is onto.*

More information on this topic can be found in [Ada74, III.8].

Homotopy inverse limits of spectra

Now suppose we have an inverse system of spectra

$$X_1 \xleftarrow{f_1} X_2 \xleftarrow{f_2} X_3 \xleftarrow{f_3} \cdots .$$

Using A.4.3 we can form the infinite product $\prod X_i$. It does *not* behave well with respect to smash products, i.e., it is not true that

$$\prod_i (X_i \wedge Y) = (\prod_i X_i) \wedge Y.$$

(This is analogous to the fact that infinite products of abelian groups do not commute with tensor products.) One can construct a map from the right hand side to the left hand side, but nothing can be proved about it in general.

This means that in general we have no way of computing the homology or cohomology of an infinite product. Two situations where we can are the following.

Proposition A.5.13 *(a) Let E and all the X_i be connective spectra and suppose that for each n, $\pi_n(X_i) = 0$ for all but finitely many i. (This occurs for example when the connectivity of X_i increases without bound as i increases.) Then*

$$E_*(\prod X_i) = \bigoplus E_*(X_i).$$

By this we mean that for each n, $E_n(\prod X_i) = \bigoplus E_n(X_i)$. For each n only finitely many $E_n(X_i)$ are nontrivial.

(b) If the spectrum E is finite then we have

$$E_*(\prod X_i) = \prod E_*(X_i).$$

for any X_i.

We can construct a shift map

$$\prod X_i \xrightarrow{\sigma} \prod X_i$$

as follows. It suffices to describe the composite

$$\prod X_i \xrightarrow{\sigma} \prod X_i \xrightarrow{p_j} X_j$$

for each $j > 0$, where p_j is the evident projection. The map we want is given by

$$p_j \sigma = p_j - f_j p_{j+1}.$$

(The minus sign refers to the additive groups structure in $[\prod X_i, X_j]$.)

Definition A.5.14 *The inverse limit of spectra $\lim_{\leftarrow} X_i$ is the fibre (i.e., the desuspension of the cofibre) of the shift map σ defined above.*

As in the case of the direct limit of spectra, this does *not* have the universal property enjoyed by the inverse limit of abelian groups, so it is not an inverse limit in the categorical sense.

From A.5.13(b) we can deduce the following.

Proposition A.5.15 *For any finite spectrum E and each integer n there is a functorial short exact sequence*

$$0 \longrightarrow \lim_{\leftarrow}{}^1 E_{n+1}(X_i) \longrightarrow E_n(\lim_{\leftarrow} X_i) \longrightarrow \lim_{\leftarrow} E_n(X_i) \longrightarrow 0.$$

In particular there is such a short exact sequence for the homotopy of an inverse limit.

This is closely related to the Milnor short exact sequence [Mil62] for the cohomology of a direct limit.

Proposition A.5.16 *For any cohomology theory E^* and each integer n there is a short exact sequence*

$$0 \longrightarrow \lim_{\leftarrow}{}^1 E^{n-1}(X_i) \longrightarrow E^n(\lim_{\rightarrow} X_i) \longrightarrow \lim_{\leftarrow} E^n(X_i) \longrightarrow 0.$$

A.6   The Adams spectral sequence

The Adams spectral sequence is a device for computing the homotopy groups of a spectrum X with the help of a ring spectrum E. Our intention here is to give a minimal introduction to it, just enough to make the references to it in the text intelligible. A much more thorough treatment and numerous references can be found in [Rav86].

The Adams spectral sequence is used three times in the text, in two very different ways. In Chapter 6 it is used to prove the periodicity theorem by constructing a map with certain properties. In this case we are using the classical Adams spectral sequence, i.e., the one based on ordinary mod p homology, in which the E_2-term can be identified with an Ext group over the Steenrod algebra. The critical point here is that these groups have a certain vanishing line, i.e., there are constants m and c such that $\mathrm{Ext}^{s,t}$ vanishes when the point (s,t) lies above the line $s = m(t - s) + c$.

It is used in a similar way in Chapter 9 to prove 9.1.6. In this case we are using a nonclassical Adams spectral sequence, i.e., one based on a generalized homology theory. The object here is to show that a certain map is nilpotent. Again the crucial point is to establish a vanishing line. This is done not by homological calculations but by the construction of an Adams resolution (A.6.1) with suitable connectivity properties.

In Chapter 8 the Adams spectral sequence is used in a very different way to prove 7.5.6 and 7.5.7. Here the homology theory being used is not connective, which means that the convergence question is very delicate. We rely heavily on some results of Bousfield [Bou79b].

The E_2-term of the Adams spectral sequence can be defined and computed in strictly algebraic terms, although these computations (especially when X is a finite complex) can be very difficult. In the most favorable cases the spectral sequence can be shown to collapse for formal reasons, i.e., there is no room for any nontrivial differentials, but failing this, one needs some additional geometric insight to compute the differentials in it. The same can be said of group extensions in the E_∞-term. The spectral sequence is known to converge to $\pi_*(X)$ under certain easily verifiable hypotheses on X and E.

Definition A.6.1 *The* **canonical Adams resolution for X based on** *E is the diagram*

$$X = X_0 \xleftarrow{\ g_0\ } X_1 \xleftarrow{\ g_1\ } X_2 \xleftarrow{\ g_2\ } \cdots$$

$$\left\downarrow{f_0}\qquad\quad \left\downarrow{f_1}\qquad\quad \left\downarrow{f_2}\right.$$

$$E \wedge X_0 \qquad E \wedge X_1 \qquad E \wedge X_2$$

where the map $X_s \to E \wedge X_s$ is induced by the unit in E, and X_{s+1} is its fibre.

More generally a (not necessarily canonical) **Adams resolution for X based on** *E is a similar diagram with $E \wedge X_s$ replaced by a spectrum K_s, such that both it and $E \wedge X_s$ are retracts of $E \wedge K_s$. X_{s+1} is still the fibre of f_s.*

See [Rav86, 2.2] for more discussion.

If we denote by $\overline{E}$ the fibre of the unit map $S^0 \to E$, in the canonical case we have

$$X_s = \overline{E}^{(s)} \wedge X.$$

Each cofibre sequence

$$X_{s+1} \xrightarrow{\ g_s\ } X_s \xrightarrow{\ f_s\ } K_s \xrightarrow{\ \partial_s\ } \Sigma X_{s+1} \tag{A.6.2}$$

gives a long exact sequence of homotopy groups. These constitute an exact couple, and a standard construction in homological algebra associates a spectral sequence with such data.

Definition A.6.3 *The* **Adams spectral sequence for** X **based on** E *is the spectral sequence based on the exact couple described above. The* E_r-*term will be denoted by* $E_r^{s,t}(X)$, *with*

$$d_r \colon E_r^{s,t}(X) \longrightarrow E_r^{s+r,t+r-1}(X).$$

$E_r^{s,t}(X)$ *vanishes for* $s < 0$, *so*

$$E_{r+1}^{s,t}(X) \subset E_r^{s,t}(X) \quad \text{for} \quad r > s$$

and we define

$$E_\infty^{s,t}(X) = \bigcap_{r>s} E_r^{s,t}(X).$$

When E *is the mod* p *Eilenberg-Mac Lane spectrum, we will sometimes refer to the Adams spectral sequence as the* **classical Adams spectral sequence.** *When* E *is* MU *or* BP, *it is often called the* **Adams-Novikov spectral sequence.**

There is a generalization of the Adams spectral sequence for computing $[W, X]_*$ rather than $\pi_*(X)$. In the E_1-term, one replaces $\pi_*(K_s)$ by $[W, K_s]_*$. For finite W it will converge whenever the Adams spectral sequence for $\pi_*(X)$ does.

Under favorable circumstances (i.e., when the Adams spectral sequence converges), $E_\infty^{s,t}(X)$ is a subquotient of $\pi_{t-s}(X)$. These subquotients are associated with the filtration of $\pi_*(X)$ defined as follows.

Definition A.6.4 *A map* $f : S^d \to X$ *has* **Adams filtration** $\geq s$ *if it can be factored as*

$$S^d \xrightarrow{f_1} W_1 \xrightarrow{f_2} W_2 \cdots \xrightarrow{f_s} X$$

where $E_*(f_i) = 0$ *for each* i. *(The spectra* W_i *for* $0 < i < s$ *are arbitrary.)*

In particular, an element has Adams filtration 0 if it is detected by homology and

$$E_\infty^{0,*} = \text{im} \left(\pi_*(x) \xrightarrow{\eta} E_*(X) \right). \tag{A.6.5}$$

Having defined the Adams spectral sequence, there are two questions we must deal with, namely

- identification of the E_2-term, and

- convergence.

The E_1 and E_2-terms

Recall (A.2.9) that the ring spectrum E is flat if $E \wedge E$ is equivalent to a wedge of various suspensions of E. All of the E we will use here, e.g. MU, BP, $X(n)$, $K(n)$ and the mod p Eilenberg-Mac Lane spectrum $H/(p)$, are flat. Being flat means that

$$E_*(E \wedge X) = E_*(E) \otimes_{\pi(E)} E_*(X).$$

When E is flat we can identify the Adams E_2-term as a certain Ext group. We will describe this first in the classical case $E = H/(p)$. The Adams E_1-term is the cochain complex

$$E_*(X_0) \xrightarrow{(f_1 \partial_0)_*} E_{*-1}(X_1) \xrightarrow{(f_2 \partial_1)_*} E_{*-2}(X_2) \xrightarrow{(f_3 \partial_2)_*} \cdots$$

$$\text{(A.6.6)}$$

The cofibre sequence of (A.6.2) induces a short exact sequence in E-homology, namely

$$0 \longrightarrow E_*(X_s) \xrightarrow{f_{s*}} E_*(E \wedge X_s) \xrightarrow{\partial_{s*}} E_{*-1}(X_{s-1}) \longrightarrow 0.$$

We can splice these together and get a long exact sequence

$$0 \longrightarrow E_*(X) \longrightarrow E_*(E \wedge X) \longrightarrow E_{*-1}(E \wedge X_1) \longrightarrow E_{*-2}(E \wedge X_2) \longrightarrow \cdots .$$

$$\text{(A.6.7)}$$

When $E = H/(p)$ we can dualize this to

$$0 \longleftarrow H^*(X) \longleftarrow H^*(E \wedge X) \longleftarrow H^{*-1}(E \wedge X_1) \longleftarrow H^{*-2}(E \wedge X_2) \longleftarrow \cdots .$$

$$\text{(A.6.8)}$$

$H^{*-1}(E \wedge X_s)$ is a free module over the Steenrod algebra A, so we have a free A-resolution of $H^*(X)$. Furthermore,

$$\pi_*(E \wedge Y) = \operatorname{Hom}_A(H^*(Y), \mathbf{Z}/(p)),$$

so applying the functor $\operatorname{Hom}_A(\cdot, \mathbf{Z}/(p))$ to (A.6.8) gives the E_1-term (A.6.6). It follows that the E_2-term, i.e., the cohomology of this cochain complex, is $\operatorname{Ext}_A(H^*(X), \mathbf{Z}/(p))$. Thus we have

Theorem A.6.9 (Adams) *If X is a connective p-local spectrum of finite type (i.e., each of its homotopy groups is finitely generated), then there is a spectral sequence converging to $\pi_*(X)$ with*

$$E_2^{s,t}(X) = \text{Ext}_A^{s,t}(H^*(X), \mathbf{Z}/(p))$$

(where the cohomology is mod p).

In order to identify the E_2-term for more general E, we need to recall the following. When E is flat, $E_*(E)$ is a Hopf algebroid (see B.3.7 below) over which $E_*(X)$ is a comodule (B.3.9). *The Adams E_2-term is Ext in the category of comodules over $E_*(E)$.* This is defined and discussed at length in [Rav86, A1.2].

Very briefly, Ext in the category of R-modules can be described in terms of derived functors of Hom. Ext in the category of comodules over a Hopf algebroid can be similarly described in terms of derived functors of the cotensor product, defined below in B.3.11.

Since the topological dimension is the number $t - s$, the Adams spectral sequence is typically depicted in a chart where the horizontal coordinate is $t - s$ and the vertical one is s. The differential d_r is a line that goes up r units while going one unit to the left.

Convergence

Given an E-based Adams resolution $\{X_s\}$ (A.6.1), let X_s' be the cofibre of the map $X_s \to X$. In particular, $X_0' = \text{pt.}$ and, in the canonical case, $X_1' = E \wedge X$. Let

$$E^{\hat{}}X = \lim_{\leftarrow} X_s', \tag{A.6.10}$$

where the inverse limit is as defined in A.5.14. Bousfield calls it the *E-nilpotent completion of X*. He shows in [Bou79b, 5.8] that it is independent of the choice of resolution. The Adams spectral sequence converges to its homotopy provided that certain $\lim^1$ groups vanish; see [Bou79b, 6.3]. Therefore *the convergence question is that of the relation between X and $E^{\hat{}}X$.*

When X is connective and E is a connective ring spectrum, then $E^{\hat{}}X$ is a functor of X depending only on the arithmetic properties of $\pi_0(E)$. Precise statements can be found in Theorems 6.5, 6.6 and 6.7 of [Bou79b]. Adams' theorem A.6.9 above is a special case.

More generally one might hope that $E^{\hat{}}X$ is the Bousfield localization $L_E X$ (7.1.1). (It follows easily from 7.1.5 and 7.1.2 that $E^{\hat{}}X$ is local.) Bousfield [Bou79b, 6.10] shows that this is the case when X is E-prenilpotent (7.1.6). He also shows in that case that the E_∞ has a horizontal vanishing line, i.e., there is an integer s_0 such that $E_\infty^{s,t} = 0$ for all $s > s_0$. This is to be expected since the E-nilpotent spectrum $L_E X$ has a finite 'E-based Postnikov system.'

A more delicate question is for which E the Adams spectral sequence always converges to $\pi_*(L_E X)$, i.e., *all* spectra are E-prenilpotent. The following result ([Bou79b, 6.12]) deals with the case when $\pi_*(E)$ is countable.

Theorem A.6.11 (Bousfield's convergence theorem) *Let E be a ring spectrum with $\pi_*(E)$ countable. Then all spectra are E-prenilpotent if and only if the E-based Adams spectral sequence satisfies the following condition: There exists a positive integer s_0 and a function φ such that for every finite spectrum X,*

$$E_\infty^{s,*}(X) = 0 \quad for \quad s > s_0$$

and

$$E_r^{s,*}(X) = E_\infty^{s,*}(X) \quad for \quad r > \varphi(s).$$

This condition says that there is a fixed horizontal vanishing line at E_∞ for all finite complexes and that the rate of convergence depends only on the filtration, not on the stem or the finite complex in question. A stronger condition would require the existence of integers $r_0, s_0 > 0$ such that for all finite X,

$$E_{r_0}^{s,*}(X) = 0 \quad for \quad s > s_0.$$

A relative form of this theorem is given in 8.2.6.

Appendix B

Complex bordism and BP-theory

In this Appendix we will outline the salient properties of complex bordism theory and BP-theory from a more technical point of view. The ideas presented here should make most of the literature on the subject intelligible.

The first two sections outline the geometric foundations of the theory, namely its connection with vector bundle and Thom spectra. In Section B.3 we take a sharp turn toward algebra and introduce Hopf algebroids, which are indispensable for the computations we need to make. The Hopf algebroid $MU_*(MU)$, the complex bordism analog of the dual Steenrod algebra, is described in Section B.4. BP-theory and the relevant Hopf algebroid, $BP_*(BP)$, are introduced in Section B.5.

In Section B.6 we discuss the Landweber exact functor theorem, a very useful tool for constructing certain new generalized homology theories. Morava K-theory is introduced in Section B.7, and in B.8 we introduce the chromatic spectral sequence and the change-of-rings-isomorphism.

B.1 Vector bundles and Thom spectra

The purpose of this section is to give the definition of the MU-spectrum, B.1.11. In order to do this we must define vector bundles (B.1.1), their classifying spaces (B.1.6), Thom spaces (B.1.8) and related constructions. The best reference for this material is Milnor-Stasheff [MS74].

Definition B.1.1 *A* **complex vector bundle** ξ *of dimension* n *over a space* B *is a map* $p: E \to B$ *such that each point* $b \in B$ *has a neighborhood*

U and a homeomorphism $h_U: p^{-1}(U) \to U \times \mathbf{C}^n$ such that the composite

$$p^{-1}(U) \xrightarrow{h_U} U \times \mathbf{C}^n \xrightarrow{p_1} U$$

(where p_1 denotes projection onto the first factor U) is the restriction of p to $p^{-1}(U)$. Moreover when two such neighborhoods U and V overlap, let h_1 and h_2 denote the restrictions of h_U and h_V to $p^{-1}(U \cap V)$. Then the composite

$$(U \cap V) \times \mathbf{C}^n \xrightarrow{h_2^{-1}} p^{-1}(U \cap V) \xrightarrow{h_1} (U \cap V) \times \mathbf{C}^n$$

has the form

$$h_1 h_2^{-1}(b, z) = g_{U,V}(b)(z)$$

for $b \in U \cap V$ and $z \in \mathbf{C}^n$, where $g_{U,V}$ is a map from $U \cap V$ to the general linear group $\mathrm{Gl}(n, \mathbf{C})$. B is the **base space** of ξ, E is its **total space**, $\mathbf{C}^n$ is its **fibre** and p is its **projection map**. The maps $g_{U,V}$ are called **transition functions**.

One can also define *real* vector bundles in a similar way. In practice it is seldom necessary to describe the transition functions explicitly.

We offer two examples of real and complex vector bundles.

Example B.1.2 (a) Let X be any space, $E = X \times \mathbf{C}^n$ and $p : E \to X$ the projection onto the first factor. The is the **trivial vector bundle over** X. The transition functions for it are all constant maps.

(b) Let $X = G_{n,k}^{\mathbf{C}}$, the space of n-dimensional complex linear subspaces of $\mathbf{C}^{n+k}$, topologized so that the set of subspaces intersecting a given open subset of $\mathbf{C}^{n+k}$ is open in $G_{n,k}^{\mathbf{C}}$. It is a smooth compact orientable manifold without boundary of dimension $2nk$ called the **Grassmannian**. (The real analogue $G_{n,k}$ is unoriented and has dimension nk.) Let

$$E = \{(x, v) \in G_{n,k}^{\mathbf{C}} \times \mathbf{C}^{n+k} : v \in x\},$$

where '$v \in x$' means that the vector v belongs to the subspace x. The map $p : E \to X$ is defined by $p(x, v) = x$. We will denote this bundle by $\gamma_{n,k}^{\mathbf{C}}$.

For the real case of B.1.2(b) with $n = k = 1$, the base space is $\mathbf{R}P^1$, the real projective line, which is homeomorphic to S^1. The total space E is homeomorphic to the interior of the Möbius strip.

Definition B.1.3 If $p: E \to Y$ is a complex vector bundle ξ over Y and $f : X \to Y$ is continuous, then the **induced bundle** $f^*(\xi)$ over X is the one with total space

$$f^*(E) = \{(x, e) \in X \times E : f(x) = p(e)\}$$

and projection map given by $p(x, e) = x$.

Definition B.1.4 *If ξ_1 and ξ_2 are vector bundles (real or complex) of dimensions n_1 and n_2 over spaces X_1 and X_2 with total spaces E_1 and E_2, then their **external sum** $\xi_1 \times \xi_2$ is a bundle of dimension $n_1 + n_2$ over $X_1 \times X_2$ given by the map*

$$E_1 \times E_2 \xrightarrow{\; p_1 \times p_2 \;} X_1 \times X_2$$

*When $X_1 = X_2 = X$, the **Whitney sum** $\xi_1 \oplus \xi_2$ is the bundle over X induced from $\xi_1 \times \xi_2$ by the diagonal map $X \to X \times X$.*

Proposition B.1.5 *(a) Let $i : G_{n,k}^{\mathbf{C}} \to G_{n,k+1}^{\mathbf{C}}$ be the map induced by the standard inclusion of $\mathbf{C}^{n+k}$ into $\mathbf{C}^{n+k+1}$, and sending an n-dimensional subspace of $\mathbf{C}^{n+k}$ to the corresponding one in $\mathbf{C}^{n+k+1}$. Then*

$$i^*(\gamma_{n,k+1}^{\mathbf{C}}) = \gamma_{n,k}^{\mathbf{C}}.$$

(b) Let $j : G_{n,k}^{\mathbf{C}} \to G_{n+1,k}^{\mathbf{C}}$ be the map induced by the standard inclusion of $\mathbf{C}^{n+k}$ into $\mathbf{C}^{n+k+1}$, and sending an n-dimensional subspace x of $\mathbf{C}^{n+k}$ to the $(n+1)$-dimensional subspace of $\mathbf{C}^{n+k+1}$ spanned by x and a fixed vector not lying in $\mathbf{C}^{n+k}$. Then

$$j^*(\gamma_{n,k+1}^{\mathbf{C}}) = \gamma_{n,k}^{\mathbf{C}} \oplus \epsilon,$$

where ϵ denotes the trivial complex line bundle.

Theorem B.1.6 *Let $BU(n)$ be the union of the spaces $G_{n,k}^{\mathbf{C}}$ under the inclusion maps i of B.1.5(a). (It is called the **classifying space for the unitary group** $U(n)$.) It is characterized (up to homotopy equivalence) by the fact that $\Omega BU(n) \simeq U(n)$.*

There is a unique n-dimensional complex vector bundle $\gamma_n^{\mathbf{C}}$ over it which induces each of the bundles $\gamma_{n,k}^{\mathbf{C}}$. This bundle is universal in the sense that any such bundle over a paracompact space X is induced by a map $X \to BU(n)$ and two such bundles over X are isomorphic if and only they are induced by homotopic maps.

*Similar statements hold in the real case. The space is called $BO(n)$, the **classifying space for the orthogonal group** $O(n)$, and its loop space is equivalent to $O(n)$. The corresponding universal bundle is denoted by γ_n.*

Theorem B.1.7 *The cohomology of $BU(n)$ is given by*

$$H^*(BU(n); \mathbf{Z}) = \mathbf{Z}[c_1, c_2, \cdots c_n]$$

*with $|c_i| = 2i$. The generator c_i is called the ith **Chern class** of the bundle $\gamma_n^{\mathbf{C}}$.*

In the real case we have

$$H^*(BO(n); \mathbf{Z}/(2)) = \mathbf{Z}/(2)[w_1, w_2, \cdots w_n]$$

with $|w_i| = i$. *The generator* w_i *is called the* i^{th} **Stiefel-Whitney class** *of the bundle* γ_n.

Under mild hypotheses on the base space X one can define an inner product on the vector space $p^{-1}(x)$ for each $x \in X$ which varies continuously with b. Such a structure is called a **Hermitian metric on** ξ.

Definition B.1.8 *Given a complex vector bundle* ξ *with a Hermitian metric over a space* B, *the* **disk bundle** $D(\xi)$ *consists of all vectors* v *with* $|v| \leq 1$ *and the* **sphere bundle** $S(\xi)$ *consists of all vectors* v *with* $|v| = 1$. *The* **Thom space** $T(\xi)$ *is the topological quotient* $D(\xi)/S(\xi)$. *(Its homeomorphism type is independent of the choice of metric.) Equivalently, if the base space* B *is compact,* $T(\xi)$ *is the one-point compactification of the total space* E. *In particular the Thom space* $T(\gamma_n^{\mathbf{C}})$ *(B.1.6) is denoted by* $MU(n)$, *and its real analog by* $MO(n)$. *A map* $f: X \to B$ *leads to a map* $T(f): T(f^*(\xi)) \to T(\xi)$ *called the* **Thomification** *of* f.

Proposition B.1.9 *(a) If the complex vector bundle* ξ *is isomorphic to the Whitney sum* $\xi' \oplus \epsilon$, *where* ϵ *denotes the trivial line bundle, then*

$$T(\xi) = \Sigma^2 T(\xi').$$

In the real case,

$$T(\xi) = \Sigma T(\xi').$$

(b) If $\xi_1 \times \xi_2$ *is an external sum (as in B.1.4) then*

$$T(\xi_1 \times \xi_2) = T(\xi_1) \wedge T(\xi_2),$$

i.e., Thomification converts external sums to smash products.

Note that if the base space B is a single point, then the Thom space is S^{2n}. Thus for each point $b \in B$ the inclusion map Thomifies to a map $S^{2n} \to T(\xi)$. There is an element $u \in H^{2n}(T(\xi); \mathbf{Z})$ (unique up to sign if B is path connected) called the **Thom class** which restricts to a generator of $H^{2n}(S^{2n})$ under each of these maps.

We remark that $D(\xi)$ is homotopy equivalent to B, so we have a relative cup product pairing (A.3.2)

$$H^*(B) \otimes H^*(D(\xi), S(\xi)) \cong H^*(D(\xi)) \otimes H^*(D(\xi), S(\xi)) \longrightarrow H^*(D(\xi), S(\xi))$$

and the latter is isomorphic to $H^*(T(\xi))$ in positive dimensions. Using this we have

Theorem B.1.10 (Thom isomorphism theorem) *(a) With notation as above, cup product with the Thom class $u \in H^{2n}(T(\xi))$ induces an isomorphism*

$$H^i(B) \xrightarrow{\Psi} H^{2n+i}(T(\xi))$$

called the **Thom isomorphism**.
(b) In the real case there is a Thom class

$$u \in H^n(T(\xi); \mathbf{Z}/(2))$$

inducing a similar isomorphism in mod 2 cohomology. An integer Thom class is defined if the bundle satisfies a certain orientability condition. This condition is always met if the base space is simply connected, or more generally if the first Stiefel-Whitney class $w_1(\xi)$ vanishes.

Now we are ready to define the Thom spectra MU and MO.

Definition B.1.11 *MU, the* **Thom spectrum for the unitary group,** *is defined by*

$$MU_{2n} = MU(n) \quad and$$
$$MU_{2n+1} = \Sigma MU(n).$$

The required map $\Sigma MU_{2n} \to MU_{2n+1}$ is the obvious one, while the map

$$\Sigma^2 MU(n) = \Sigma MU_{2n+1} \longrightarrow MU_{2n+2} = MU(n+1)$$

is as follows. There is a map $j : BU(n) \to BU(n+1)$ induced by the maps $j : G^{\mathbf{C}}_{n,k} \to G^{\mathbf{C}}_{n+1,k}$ of B.1.5(b), with

$$j^*(\gamma^{\mathbf{C}}_{n+1}) = \gamma^{\mathbf{C}}_n \oplus \epsilon.$$

Hence the Thomification of j is (by B.1.9) the desired map

$$\Sigma^2 MU(n) \longrightarrow MU(n+1).$$

In the real case, MO, the **Thom spectrum for the orthogonal group,** *is given by*

$$MO_n = MO(n).$$

Definition B.1.12 *Given a map $f : Y \to BU$, the* **associated Thom spectrum** *X is defined as follows. The restriction of f to the 2n-skeleton Y^{2n} of Y factors uniquely through $BU(n)$ and hence (B.1.6) defines an n-dimensional complex vector bundle over Y^{2n}. X_{2n} is its Thom space and $X_{2n+1} = \Sigma X_{2n}$.*
The Thom spectrum associated with a map $Y \to BO$ can be defined similarly.

Note that MU and MO as we have defined them here are neither suspension spectra nor Ω-spectra, although they are equivalent to the latter by (A.2.4).

We still need to explain the multiplicative structure on MU. We need a map

$$MU \wedge MU \xrightarrow{m} MU$$

satisfying the conditions of A.2.8. In this case the naive definition (5.1.3) of the smash product of two spectra is sufficient, because we will produce maps of spaces

$$MU(n_1) \wedge MU(n_2) \xrightarrow{m} MU(n_1 + n_2).$$

We have a map

$$G^{\mathbf{C}}_{n_1, k_1} \times G^{\mathbf{C}}_{n_2, k_2} \xrightarrow{m} G^{\mathbf{C}}_{n_1 + n_2, k_1 + k_2}$$

which sends a pair of subspaces to their direct sum, and

$$m^*(\gamma^{\mathbf{C}}_{n_1 + n_2, k_1 + k_2}) = \gamma^{\mathbf{C}}_{n_1, k_1} \times \gamma^{\mathbf{C}}_{n_2, k_2}.$$

Letting k_1 and k_2 go to ∞ we get

$$BU(n_1) \times BU(n_2) \xrightarrow{m} BU(n_1 + n_2)$$

with

$$m^*(\gamma^{\mathbf{C}}_{n_1 + n_2}) = \gamma^{\mathbf{C}}_{n_1} \times \gamma^{\mathbf{C}}_{n_2}.$$

The induced map in cohomology (see B.1.7) is the *Whitney sum formula*,

$$m^*(c_i) = \sum_{j+k=i} c_j \otimes c_k \tag{B.1.13}$$

for $0 \le i \le n_1 + n_2$, where it is understood that $c_0 = 1$, and $c_j \otimes c_k = 0$ if $j > n_1$ or $k > n_2$. This can be rewritten as

$$m^*(\sum_{i \ge 0} c_i) = \sum_{j \ge 0} c_j \otimes \sum_{k \ge 0} c_k.$$

There is a similar formula in the real case involving Stiefel-Whitney classes.

By B.1.9, m Thomifies to a map

$$MU(n_1) \wedge MU(n_2) \xrightarrow{\;\;T(m)\;\;} MU(n_1 + n_2).$$

These are the maps needed to make MU a ring spectrum.

We can use the Whitney sum formula (B.1.13) to compute $H_*(MU)$ as a ring as follows. It tells us that the map $H^*(m)$ is monomorphic, so $H_*(m)$ is onto. The same goes for the iterated Whitney sum map

$$BU(1) \times BU(1) \times \cdots BU(1) \xrightarrow{m(n)} BU(n)$$

with n factors in the source. We know that

$$G^{\mathbf{C}}_{1,n+1} = \mathbf{C}P^n \text{ (complex projective } n\text{-space), so}$$
$$BU(1) = \mathbf{C}P^\infty \text{ and}$$
$$H^*(BU(1)^n) = \mathbf{Z}[x_1, x_2, \cdots x_n]$$

with $|x_i| = 2$.

(B.1.13) implies that $m(n)^*(c_j)$ is the jth elementary symmetric function in the x_i, i.e.,

$$m(n)^*(\sum_{j=0}^{n} c_j) = \prod_{i=1}^{n}(1 + x_i).$$

To describe the situation in homology, let

$$b_i \in H_{2i}(BU(1)) \qquad (B.1.14)$$

be the generator dual to c_1^i. Equivalently, b_i is the image of the fundamental homology class in $\mathbf{C}P^i$ under the usual inclusion map.

We will denote the class

$$m(n)_*(b_{i_1} \otimes b_{i_2} \otimes \cdots b_{i_n}) \in H_*(BU(n))$$

by $b_{i_1} b_{i_2} \cdots b_{i_n}$. A monomial of degree $m < n$ will denote the image of the corresponding class in $H_*(BU(m))$. This 'multiplication' is commutative due to the symmetry in the Whitney sum formula (B.1.13).

This notation is slightly misleading in that there is no natural multiplication in $H_*(BU(n))$, i.e. there is no map

$$BU(n) \times BU(n) \longrightarrow BU(n)$$

with suitable properties. However there is a natural multiplication in $H_*(MU)$. From these considerations and B.1.10 we can deduce the following.

Theorem B.1.15 *The homology of the spectrum MU is given by*

$$H_*(MU) = \mathbf{Z}[b_1, b_2, \cdots]$$

with $|b_i| = 2i$. Moreover, the image of $H_*(MU(n))$ is the subgroup spanned by all monomials of degree $\leq n$.

This notation is slightly abusive for the following reason. The generator $b_i \in H_{2i}(MU)$ pulls back to $H_{2i+2}MU(1)$. An easy geometric exercise shows that

$$T(\gamma_{1,k}^{\mathbf{C}}) = \mathbf{C}P^{k+1}$$

so

$$MU(1) = \mathbf{C}P^{\infty}.$$

The class here that maps to $b_i \in H_{2i}(MU)$ is $b_{i+1} \in H_{2i+2}(\mathbf{C}P^{\infty})$.

B.2   The Pontrjagin-Thom construction

In this section we will explain the relation between MU and cobordism of complex manifolds. While this connection was important historically, from the point of view of this book it is merely an amusing diversion. The results stated in this section are not needed elsewhere. Again we refer the reader to [MS74] for a more detailed account of what follows. Further information on cobordism theory can be found in Stong's book [Sto68].

Let M^k be a compact smooth k-dimensional manifold without boundary smoothly embedded in the Euclidean space $\mathbf{R}^{2n+k}$. The **tubular neighborhood theorem** asserts that M has an open neighborhood $V \subset \mathbf{R}^{2n+k}$ with a map $p: V \to M$ (which is the identity on M itself) which is a $2n$-dimensional real vector bundle over M called its **normal bundle** ν. One has a disk bundle $D(\nu) \subset V$, which is a $(2n + k)$-dimensional manifold whose boundary is the sphere bundle $S(\nu)$. We get a map $\mathbf{R}^{2n+k} \to T(\nu)$ by sending everything outside of $D(\nu)$ to the base-point (i.e., the image of $S(\nu)$) in $T(\nu)$. This map extends to S^{2n+k}, the one-point compactification of $\mathbf{R}^{2n+k}$.

Suppose that this bundle admits a complex linear structure, i.e., it is an n-dimensional complex vector bundle. Then ν is induced by a map $M \to BU(n)$, which Thomifies to a map $T(\nu) \to MU(n)$. The composition

$$S^{2n+k} \longrightarrow T(\nu) \longrightarrow MU(n)$$

is denoted by Φ_M, called the **Pontrjagin-Thom construction on** M. (Strictly speaking, it is determined not just by M but by the embedding of M in $\mathbf{R}^{2n+k}$.) Note that we can recover M from Φ_M as the preimage of $BU(n) \subset MU(n)$.

Now we wish to define an equivalence relation on embedded manifolds which translates to homotopy under this Φ.

Definition B.2.1 *Two compact smooth k-dimensional manifolds without boundary smoothly embedded in the Euclidean space $\mathbf{R}^{2n+k}$, M_1 and M_2 are **cobordant** if there is a compact smooth $(k + 1)$-dimensional manifold*

$$W \subset \mathbf{R}^{2n+k} \times [0,1]$$

such that

$$W \cap \mathbf{R}^{2n+k} \times \{0\} = M_1$$

and

$$W \cap \mathbf{R}^{2n+k} \times \{1\} = M_2.$$

W is called a **cobordism** *between M_1 and M_2. It is a* **complex cobordism** *if its normal bundle in $\mathbf{R}^{2n+k} \times [0,1]$ admits a complex linear structure.*

This definition should be compared with 3.1.1; the difference is that there the maps are not required to be embeddings. Cobordism as defined in B.2.1 is also an equivalence relation and the set of cobordism classes forms a group under disjoint union, which we denote by $\Omega_{k,n}^U$.

We can apply the Pontrjagin-Thom construction to W and obtain a map

$$\Phi_W : S^{n+k} \times [0,1] \longrightarrow MU(n)$$

which is a homotopy between Φ_{M_1} and Φ_{M_2} Thus we have a homomorphism

$$\Omega_{k,n}^U \xrightarrow{\ \Phi\ } \pi_{k+2n}(MU(n)).$$

The **Thom transversality theorem** asserts that any map $S^{2n+k} \to MU(n)$ can be approximated by (and is therefore homotopic to) one obtained by the Pontrjagin-Thom construction. This implies that Φ is onto. A similar argument shows that every homotopy between maps given by the Pontrjagin-Thom construction can be approximated by one coming from a cobordism, which shows that Φ is one-to-one. Hence we have

Theorem B.2.2 (Thom cobordism theorem) *The Pontrjagin-Thom homomorphism*

$$\Phi : \Omega_{k,n}^U \to \pi_{2n+k}(MU(n))$$

above is an isomorphism.

This theorem translates the geometric problem of determining the cobordism group $\Omega_{k,n}^U$ into that of computing the homotopy group $\pi_{2n+k}(MU(n))$. Both groups are known to be independent of n if n is large enough (roughly if $2n > k$). For large n the cobordism group is denoted simply by Ω_k^U. These groups form a graded ring under Cartesian product of manifolds.

The homotopy groups for large n are the homotopy groups of the spectrum MU defined by $MU_{2n} = MU(n)$ and $MU_{2n+1} = \Sigma MU(n)$. Indeed it was this type of example that motivated the definition of spectra (5.1.1) in the first place. $\pi_*(MU)$ will be described below in B.4.1.

B.3   Hopf algebroids

In this section we will give the definition of a Hopf algebroid B.3.7, of which $MU_*(MU)$ is the motivating example. This is a generalization of the definition of a Hopf algebra, of which the dual Steenrod algebra A_* is a classical example. Much more information on this topic can be found in [Rav86, A1].

In order to motivate the definition we need to recall some classical ideas from algebraic topology. For more information, see [Rav86, 2.2] and [Ada74, III.12].

Let E be H/p, the mod p Eilenberg-Mac Lane spectrum, so $E^*(X) = H^*(X; \mathbf{Z}/(p))$. This is a graded $\mathbf{Z}/(p)$-vector space whose structure is enriched by the action of certain natural operations called *Steenrod operations*. These form the *Steenrod algebra* A over which $H^*(X; \mathbf{Z}/(p))$ has a natural module structure. In our notation this structure is a map

$$A \otimes E^*(X) \longrightarrow E^*(X)$$

with certain properties. Taking the $\mathbf{Z}/(p)$-linear dual we get a homomorphism

$$A_* \otimes E_*(X) \xleftarrow{\ \psi\ } E_*(X) \tag{B.3.1}$$

with similar properties. Here A_* denotes the linear dual of the Steenrod algebra. We call ψ a *comodule structure* on $E_*(X)$.

The Steenrod algebra A can be identified with the mod p cohomology of the mod p Eilenberg-Mac Lane spectrum E, i.e., with $E^*(E)$. It follows that its dual A_* can be identified with $E_*(E)$ and we can rewrite (B.3.1) as

$$\pi_*(E \wedge E \wedge X) \xleftarrow{\ \psi\ } \pi_*(E \wedge X). \tag{B.3.2}$$

It can be shown that this is induced by the map

$$E \wedge E \wedge X \xleftarrow{\quad E \wedge \eta \wedge X \quad} E \wedge S^0 \wedge X = E \wedge X$$

where $\eta : S^0 \to E$ is the unit map for the ring spectrum E.

The homomorphism ψ of (B.3.2) can be defined for any ring spectrum E, including MU. We have an isomorphism

$$\pi_*(E \wedge E \wedge X) = E_*(E) \otimes_{E_*} E_*(X)$$

under the assumption that E is flat in the sense of A.2.9. Both H/p and MU are flat, while H (the integer Eilenberg-Mac Lane spectrum) is not.

Now we will describe the formal properties of $E_*(E)$ the and map ψ of (B.3.2) for a flat ring spectrum (A.2.9) E. In the classical case (i.e., $E = H/p$), $E_*(E)$, the dual Steenrod algebra A_*, is a commutative Hopf algebra. This means it has structure maps

$$
\begin{array}{rcll}
A_* & \xrightarrow{\Delta} & A_* \otimes A_* & \text{(coproduct)} \\
A_* \otimes A_* & \xrightarrow{\mu} & A_* & \text{(product)} \\
A_* & \xrightarrow{c} & A_* & \text{(conjugation)} \\
A_* & \xrightarrow{\epsilon} & \mathbf{Z}/(p) & \text{(augmentation)} \\
\mathbf{Z}/(p) & \xrightarrow{\eta} & A_* & \text{(unit)}
\end{array}
\qquad \text{(B.3.3)}
$$

satisfying certain conditions.

The most concise way to specify these conditions is to say that A_* is *a cogroup object in the category of graded commutative $\mathbf{Z}/(p)$-algebras with unit.* This means that for any other such algebra C, there is a natural group structure on the set $\mathrm{Hom}(A_*, C)$ of graded algebra homomorphisms. In particular, the coproduct Δ on A_* induces a map

$$
\mathrm{Hom}(A_*, C) \times \mathrm{Hom}(A_*, C) = \mathrm{Hom}(A_* \otimes A_*, C) \xrightarrow{\Delta^*} \mathrm{Hom}(A_*, C)
$$

which is the group operation on $\mathrm{Hom}(A_*, C)$. The conjugation c induces the map that sends each element to its inverse and the augmentation ϵ gives the identity element. The other two maps, the product μ and the unit η, define A_* itself as a graded commutative algebra with unit. With this in mind, the defining properties of a group can be translated into the properties of the structure maps that define a graded commutative Hopf algebra such as A_*.

The following explicit description of the dual Steenrod algebra A_* is originally due to Milnor [Mil58].

Theorem B.3.4 *For $p = 2$ the dual Steenrod algebra is*

$$
A_* = \mathbf{Z}/(2)[\xi_1, \xi_2, \cdots]
$$

with $|\xi_n| = 2^n - 1$ and the coproduct is given by

$$
\Delta(\xi_n) = \sum_{i=0}^{n} \xi_{n-i}^{2^i} \otimes \xi_i
$$

where $\xi_0 = 1$.
For $p > 2$,

$$
A_* = \mathbf{Z}/(p)[\xi_1, \xi_2, \cdots] \otimes E(\tau_0, \tau_1, \cdots)
$$

with $|\xi_n| = 2p^n - 2$ and $|\tau_n| = 2p^n - 1$. The coproduct is given by

$$\Delta(\xi_n) = \sum_{i=0}^{n} \xi_{n-i}^{p^i} \otimes \xi_i$$

$$\Delta(\tau_n) = \tau_n \otimes 1 + \sum_{i=0}^{n} \xi_{n-i}^{p^i} \otimes \tau_i.$$

All but the last of the five maps in (B.3.3) has a topological counterpart. If we identify $\mathbf{Z}/(p)$ with $\pi_*(E)$, A_* with $\pi_*(E \wedge E)$ and $A_* \otimes A_*$ with $\pi_*(E \wedge E \wedge E)$, then the first four of the maps are induced by

$$
\begin{array}{llll}
E \wedge E = E \wedge S^0 \wedge E & \xrightarrow{\ E \wedge \eta \wedge E\ } & E \wedge E \wedge E & \text{(coproduct)} \\
E \wedge E \wedge E & \xrightarrow{\ E \wedge m\ } & E \wedge E & \text{(product)} \\
E \wedge E & \xrightarrow{\ T\ } & E \wedge E & \text{(conjugation)} \\
E \wedge E & \xrightarrow{\ m\ } & E & \text{(augmentation)},
\end{array}
$$

where η and m are the unit and multiplication maps for the ring spectrum E, and T is the map which interchanges the two factors.

This leaves the unit map $\eta : \pi_*(E) \to \pi_*(E \wedge E)$. For this there are two natural choices, namely

$$
\begin{array}{lll}
E = S^0 \wedge E & \xrightarrow{\ \eta \wedge E\ } & E \wedge E \quad \text{and} \\
E = E \wedge S^0 & \xrightarrow{\ E \wedge \eta\ } & E \wedge E.
\end{array}
$$

When $E = H/p$, these two maps induce the same homomorphism in homotopy, namely the unit map of (B.3.3), but in general they will induce different maps, which are denoted by η_R and η_L, the right and left units respectively.

For a flat ring spectrum E these two units figure in the isomorphism

$$\pi_*(E \wedge E \wedge E) = E_*(E) \otimes_{E_*} E_*(E) \tag{B.3.5}$$

as follows. η_R and η_L determine two different E_*-module structures on $E_*(E)$, called the right and left module structures. In other words, they make $E_*(E)$ a two-sided E_*-module. This structure is used to define the tensor product in (B.3.5).

Thus in order to describe $MU_*(MU)$ we need to generalize the notion of a graded commutative Hopf algebra in such a way as to accommodate the presence of distinct right and left units. We will do this by generalizing the notion of a group object in a category.

For this we need some more category theory. A group can be thought of as a category with one object in which every morphism is invertible. The morphisms in this category correspond to the elements in the group, and composition of morphisms corresponds to multiplication.

With this in mind, a *groupoid* is a small category (i.e., a category in which the collection of objects is a set rather than a class) in which every morphism is invertible. It differs from a group in that the category is allowed to have more than one object. There are two sets associated with a groupoid, namely the set $\mathbf{O}$ of objects and the set $\mathbf{M}$ of morphisms. We will regard a groupoid as a pair of sets $(\mathbf{O}, \mathbf{M})$ endowed with the following structure.

There is a map $\mathbf{O} \to \mathbf{M}$ which assigns to each object its identity morphism, a map $\mathbf{M} \to \mathbf{M}$ which assigns to each morphism its inverse, and two maps $\mathbf{M} \to \mathbf{O}$, assigning to each morphism its source and target. Composition is a map $\mathbf{D} \to \mathbf{M}$, where

$$\mathbf{D} = \{(m_1, m_2) \in \mathbf{M} \times \mathbf{M} : \text{target}(m_1) = \text{source}(m_2)\} \qquad \text{(B.3.6)}$$

These maps must satisfy certain conditions which the interested reader can easily spell out.

A general reference for groupoids is the book by Higgins [Hig71].

Definition B.3.7 *A **Hopf algebroid** over K is a pair (S, Σ) of graded commutative K-algebras with unit which is a cogroupoid object in the category of such algebras, i.e., given any such algebra C, the pair of sets*

$$(\text{Hom}(S, C), \text{Hom}(\Sigma, C))$$

has a natural groupoid structure as described above. In other words S and Σ are endowed with structure maps making $\text{Hom}(S, C)$ and $\text{Hom}(\Sigma, C)$ the object and morphism sets of a groupoid. More explicitly we have

$$
\begin{array}{lll}
\Sigma \xrightarrow{\;\;\Delta\;\;} \Sigma \otimes_S \Sigma & & \textit{(coproduct inducing composition)} \\
\Sigma \xrightarrow{\;\;c\;\;} \Sigma & & \textit{(conjugation inducing inverse)} \\
\Sigma \xrightarrow{\;\;\epsilon\;\;} S & & \textit{(augmentation inducing identity morphism)} \\
S \xrightarrow{\;\;\eta_R\;\;} \Sigma & & \textit{(right unit inducing target)} \\
S \xrightarrow{\;\;\eta_L\;\;} \Sigma & & \textit{(left unit inducing source)}
\end{array}
$$

satisfying conditions corresponding to the groupoid structure. As before, the tensor product

$$\Sigma \otimes_S \Sigma$$

is defined in terms of the right and left units.

To see why $\Sigma \otimes_S \Sigma$ is the appropriate target for Δ, recall that composition of morphisms in a groupoid $(\mathbf{O}, \mathbf{M})$ is a map

$$\mathbf{M} \times \mathbf{M} \supset \mathbf{D} \longrightarrow \mathbf{M}$$

where the subset $\mathbf{D} \subset \mathbf{M} \times \mathbf{M}$ is as in (B.3.6). This amounts to saying that it is the difference kernel of the two maps

$$\mathbf{M} \times \mathbf{M} \; \overline{} \; \mathbf{M} \times \mathbf{O} \times \mathbf{M}$$

where the two maps send (m_1, m_2) to

$$(m_1, \text{target}(m_1), m_2) \qquad \text{and}$$
$$(m_1, \text{source}(m_2), m_2).$$

We need the dual to this construction. It is the difference cokernel of

$$\Sigma \otimes \Sigma \; \overline{} \; \Sigma \otimes S \otimes \Sigma$$

where the two maps send $\sigma_1 \otimes s \otimes \sigma_2$ to

$$(\sigma_1 \eta_R(s), \, \sigma_2) \qquad \text{and}$$
$$(\sigma_1, \, \eta_L(s)\sigma_2).$$

This cokernel is $\Sigma \otimes_S \Sigma$ by definition.

The relevance of Hopf algebroids to homotopy theory is the following.

Theorem B.3.8 *If E is a flat (A.2.9) homotopy commutative ring spectrum, then $(\pi_*(E), E_*(E))$ is a Hopf algebroid over $\mathbf{Z}$. If E is p-local, it is a Hopf algebroid over $\mathbf{Z}_{(p)}$.*

We also need a generalization of the comodule structure of (B.3.1).

Definition B.3.9 *A left comodule over a Hopf algebroid (S, Σ) is a left S-module M together with a left S-linear map*

$$M \xrightarrow{\psi} \Sigma \otimes_S M$$

which is counitary and coassociative, i.e., the composite

$$M \xrightarrow{\psi} \Sigma \otimes_S M \xrightarrow{\epsilon \otimes M} S \otimes_S M = M$$

is the identity on M and the diagram

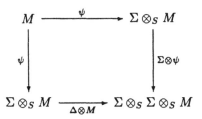

commutes.

A **right comodule** *is defined similarly.*

A **comodule algebra** *is an S-algebra A which is a comodule such that the map ψ is an algebra homomorphism.*

Proposition B.3.10 *If E is as in B.3.8 then for any X, $E_*(X)$ is a left comodule over $E_*(E)$ with the homomorphism ψ induced by the map*

$$E \wedge X = E \wedge S^0 \wedge X \xrightarrow{\ E \wedge \eta \wedge X\ } E \wedge E \wedge X.$$

The homological algebra associated with Hopf algebroids is discussed at length in [Rav86, A1.2]. In particular one can define Ext groups in terms of derived functors of the cotensor product, defined as follows.

Definition B.3.11 *Let M be a right comodule over the Hopf algebroid (S, Σ) with structure map $\phi : M \to M \otimes_S \Sigma$ and N a left comodule with structure map $\psi : N \to \Sigma \otimes_S N$. Then the **cotensor product** $M \square_\Sigma N$ is the kernel of the map*

$$M \otimes_S N \xrightarrow{\ \phi \otimes 1_N - 1_M \otimes \psi\ } M \otimes_S \Sigma \otimes_S N$$

B.4   The structure of $MU_*(MU)$

In this section we will describe the Hopf algebroid (B.3.7)

$$(\pi_*(MU), MU_*(MU))$$

explicitly. More information can be found in [Rav86, 4.1] and [Ada74, II.11].

This Hopf algebroid can be described in two different ways. One can give explicit formulae for its structure maps, and one can define the groupoid-valued functor represented by it. The value of this functor on a graded commutative algebra R is, roughly speaking, the set of formal group laws over R and the groupoid of isomorphisms between them.

$\pi_*(MU)$ is the ring L of 3.2.3. Thus $MU_*(X)$ is an L-module. We will explain why the coaction over $MU_*(MU)$ is equivalent to an action of the group Γ of 3.3.1, so the category of comodules over $MU_*(MU)$ is equivalent to the category $C\Gamma$ of 3.3.

While the notion of a Γ-action may be more conceptually convenient, a coaction of $MU_*(MU)$ is more amenable to computation. For this reason it is the language used in almost all of the literature on the subject. For p-local computations, the language of BP-theory is still more convenient. It will be discussed below in B.5.

$\pi_*(MU)$ was computed independently by Milnor [Mil60] and Novikov ([Nov60] and [Nov62]) using the Adams spectral sequence, which we introduced here in A.6. A proof is given in [Rav86, Theorem 3.1.5]. Their result is

Theorem B.4.1 (Milnor-Novikov theorem) *Let MU be the Thom spectrum associated with the unitary group U as described above.*

(a)
$$\pi_*(MU) \cong \mathbf{Z}[x_1, x_2, \ldots]$$
where $\dim x_i = 2i$, *and*

(b) the generators x_i (which we shall not define here) can be chosen in such a way that the Hurewicz map $h: \pi_*(MU) \to H_*(MU)$ (see B.1.15) is given by

$$h(x_i) = \begin{cases} pb_i + decomposables & if\ i = p^k - 1\ for\ some\ prime\ p \\ b_i + decomposables & otherwise. \end{cases}$$

The structure of $MU_*(MU)$ is originally due to Landweber [Lan67] and Novikov [Nov67]. In order to state it we need the isomorphism of B.2.2 between the complex cobordism ring and $\pi_*(MU)$. With that in mind, let

$$m_n = \frac{[\mathbf{C}P^n]}{n+1} \in \pi_{2n}(MU) \otimes \mathbf{Q} \tag{B.4.2}$$

where $[\mathbf{C}P^n]$ is the cobordism class represented by complex projective space $\mathbf{C}P^n$.

Theorem B.4.3 (Landweber-Novikov theorem) *As a ring,*

$$MU_*(MU) = MU_*[b_1, b_2, \cdots]$$

with $|b_i| = 2i$.

The coproduct is given by

$$\sum_{i \geq 0} \Delta(b_i) = \sum_{i \geq 0} b_i \otimes \left(\sum_{j \geq 0} b_j \right)^{i+1}$$

where $b_0 = 1$. (This means that $\Delta(b_n)$ can be found by expanding the right hand side and taking the terms in dimension $2n$, of which there are only a finite number.)

The left unit η_L is the standard inclusion

$$MU_* \longrightarrow MU_*(MU) = MU_*[b_1, b_2, \cdots]$$

while the right unit on $MU_*(MU) \otimes \mathbf{Q}$ is given by

$$\sum_{i \geq 0} \eta_R(m_i) = \sum_{i \geq 0} m_i \left(\sum_{j \geq 0} c(b_j) \right)^{i+1}$$

where $m_0 = 1$ and c is the conjugation.

The conjugation c is given by $c(m_n) = \eta_R(m_n)$ and

$$\sum_{i \geq 0} c(b_i) \left(\sum_{j \geq 0} b_j \right)^{i+1} = 1.$$

At first glance the above appears to be a description of $MU_*(MU) \otimes \mathbf{Q}$ rather than $MU_*(MU)$ itself. The theorem is meant to assert that given conjugation and right unit on $MU_*(MU) \otimes \mathbf{Q}$ induce ones on $MU_*(MU)$.

The fact that the coproduct Δ sends each b_n to a polynomial in the $b_i \otimes b_j$ with coefficients in the ground ring $\mathbf{Z}$ rather than the object ring MU_* is a special property of this particular Hopf algebroid. It means that $B = \mathbf{Z}[b_1, b_2, \cdots]$ is a Hopf algebra over $\mathbf{Z}$ and the Hopf algebroid structure is determined by B along with the right unit map, which amounts to a B-comodule structure on MU_*.

The term for this property is *split*. It corresponds to the following property of groupoids.

Definition B.4.4 *A groupoid is* **split** *if it is obtained in the following way. Let X be a set acted on by a group G. Regard the elements of X as objects in a category* $\mathbf{C}$ *whose morphism set is* $G \times X$. *The source and target of the morphism* (g, x) *are* x *and* $g(x)$, *and the composite of* (g, x) *and* $(g', g(x))$ *is* $(g'g, x)$.

Definition B.4.5 *A Hopf algebroid* (S, Σ) *over* K *is* **split** *if* $\Sigma = S \otimes B$ *where* B *is a Hopf algebra over which* S *is a comodule algebra (B.3.9).*

Note that when (S, Σ) is split, then for any graded commutative K-algebra C,

$$\mathrm{Hom}(\Sigma, C) = \mathrm{Hom}(S, C) \times \mathrm{Hom}(B, C)$$

and $\text{Hom}(B, C)$ is a group (since B is a Hopf algebra) acting on the set $\text{Hom}(S, C)$.

Now consider the case at hand. For a moment we will ignore the gradings on our algebras. Let R be a commutative ring. Then by the theorems of Lazard 3.2.3 and Quillen 3.2.4 we can identify the set

$$\text{Hom}(MU_*, R) \qquad (B.4.6)$$

with the set of formal group laws over R, which we will denote by $\text{FGL}(R)$.

The Hopf algebra B is a ring of integer valued algebraic functions on Γ. To see this, write an element $\gamma \in \Gamma$ as

$$\gamma = \sum_{i \geq 0} b_i x^{i+1}$$

with $b_0 = 1$. Then we can regard b_i as the function which assigns to γ the coefficient of x^{i+1} in its power series expansion.

Proposition B.4.7 *Let B be the Hopf algebra $\mathbf{Z}[b_1, b_2, \cdots]$ with the co-product given in B.4.3. Then for a commutative ring R, $\text{Hom}(B, R)$ is the group Γ_R of power series of the form*

$$x + b_1 x^2 + b_2 x^3 + \cdots$$

with coefficients over R under functional composition.

Proof. The notation is meant to indicate that the power series associated with a homomorphism $f : B \to R$ is $\sum_{i \geq 0} f(b_i) x^{i+1}$ with $b_0 = 1$. We need only to check that the coproduct in B given in B.4.3 induces the appropriate group structure. Suppose

$$\sum_{i \geq 0} b'_i x^{i+1} \qquad \text{and} \qquad \sum_{j \geq 0} b''_j x^{j+1}$$

are two such power series. Then their composite is

$$\sum_{i \geq 0} b'_i (\sum_{j \geq 0} b''_j x^{j+1})^{i+1}$$

and this corresponds precisely to the coproduct in B. ∎

The action of the group Γ_R on the set $\text{FGL}(R)$ is as follows. Given a formal group law F over R and a power series

$$f(x) = \sum_{i \geq 0} b_i x^{i+1}$$

with $b_0 = 1$ and $b_i \in R$, it is easy to verify that the power series

$$F^f(x, y) = f(F(f^{-1}(x), f^{-1}(y))) \tag{B.4.8}$$

is another formal group law over R.

In order to interpret the right unit in $MU_*(MU)$ in this light, we need some more of the theory of formal group laws. Let $F(x, y)$ be a formal group law over a torsion free ring R. Let

$$F_2(x, y) = \frac{\partial F}{\partial y}$$

and define the **logarithm** of F by the formula

$$\log_F(x) = \int_0^x \frac{dt}{F_2(t, 0)}. \tag{B.4.9}$$

This is a power series over $R \otimes \mathbf{Q}$ since we must be able to divide by exponents in order to integrate.

For example, for the multiplicative formal group law of 3.2.2 we have

$$
\begin{aligned}
F(x, y) &= x + y + xy \\
F_2(x, y) &= 1 + x \\
\log_F(x) &= \int_0^x \frac{dt}{1 + t}
\end{aligned}
$$

We tell our calculus students that this is $\log(1 + x)$, hence the notation $\log_F(x)$. We recommend calculating it for other examples 3.2.2 as an exercise for the reader.

Proposition B.4.10 *For any formal group law over a torsion free ring R,*

$$\log_F(F(x, y)) = \log_F(x) + \log_F(y).$$

Proof. Let

$$w(x, y) = \log_F(F(x, y)) - \log_F(x) + \log_F(y).$$

In a moment we will show that

$$\frac{\partial w}{\partial y} = 0. \tag{B.4.11}$$

Since w is symmetric in x and y it will follow that

$$\frac{\partial w}{\partial x} = 0$$

and therefore $w(x, y)$ is constant. It is clear from its definition that the constant term in $\log_F(x)$ is zero, so if $w(x, y)$ is constant, $w(x, y) = 0$ and the result follows.

Before we can prove (B.4.11), we need to consider the associativity condition

$$F(F(x, y), z) = F(x, F(y, z)).$$

Differentiating with respect to z and then setting $z = 0$ gives

$$F_2(F(x, y), 0) = F_2(x, y)F_2(y, 0). \tag{B.4.12}$$

Now we are ready to complete the proof by showing (B.4.11). Using the fundamental theorem of calculus we have

$$
\begin{aligned}
\frac{\partial w}{\partial y} &= \frac{\partial}{\partial y}\left(\int_0^{F(x,y)} \frac{dt}{F_2(t,0)} - \int_0^x \frac{dt}{F_2(t,0)} - \int_0^y \frac{dt}{F_2(t,0)} \right) \\
&= \frac{F_2(x, y)}{F_2(F(x, y), 0)} - \frac{1}{F_2(y, 0)} \\
&= \frac{F_2(x, y)F_2(y, 0) - F_2(F(x, y), 0)}{F_2(F(x, y), 0)F_2(y, 0)}
\end{aligned}
$$

and this vanishes by (B.4.12).   ∎

B.4.10 says that $\log_F(x)$ is an isomorphism between an arbitrary formal group law F over a $\mathbf{Q}$-algebra and the additive formal group law. Since Lazard's ring L is torsion free, the logarithm provides a convenient way of doing computations with it. A theorem of Mischenko says that for the universal formal group law $G(x, y)$ of 3.2.3,

$$\log_G(x) = \sum_{i \geq 0} m_i x^{i+1}$$

where m_i is as in (B.4.2).

Now we want to compute the logarithm of the formal group law F^f of (B.4.8) in terms of that of F. We can rewrite (B.4.8) as

$$f^{-1}(F^f(x, y)) = F(f^{-1}(x), f^{-1}(y)).$$

Applying $\log_F$ to both sides we get

$$
\begin{aligned}
\log_F(f^{-1}(F^f(x,y))) &= \log_F(F(f^{-1}(x), f^{-1}(y))) \\
&= \log_F(f^{-1}(x)) + \log_F(f^{-1}(y))
\end{aligned}
$$

We also have

$$\log_{F^f}(F^f(x,y)) = \log_{F^f}(x) + \log_{F^f}(y)$$

Comparing these two and arguing by induction on degree we conclude that

$$\log_{F^f}(x) = \log_F(f^{-1}(x))$$

(Compare with the proof of [Rav86, A2.1.16], which contains an unfortunate misprint. The last displayed equation in the proof should read $mog(x) = \log(f^{-1}(x))$.)

Now given a homomorphism $\theta : MU_*(MU) \to R$, we have

$$
\begin{array}{ccc}
 & B & \\
 & \downarrow & \\
MU_* \underset{\eta_R}{\overset{\eta_L}{\rightrightarrows}} MU_*(MU) & \overset{\theta}{\longrightarrow} & R
\end{array}
$$

The composite map $B \to R$ gives the coefficients of the power series $f(x)$, $\theta\eta_L$ gives the formal group law F and $\theta\eta_R$ gives the formal group law F^f. Hence

$$
\begin{aligned}
\log_F(x) &= \sum_{i\geq 0} \theta(m_i)x^{i+1} \qquad \text{and} \\
\log_{F^f}(x) &= \sum_{i\geq 0} \theta(\eta_R(m_i))x^{i+1} \\
&= \sum_{i\geq 0} \theta(m_i)(f^{-1}(x))^{i+1} \\
&= \sum_{i\geq 0} \theta(m_i)(\sum_{j\geq 0} \theta(c(b_j))x^{j+1})^{i+1} \qquad \text{since} \\
f^{-1}(x) &= \sum_{j\geq 0} \theta(c(b_j))x^{j+1}.
\end{aligned}
$$

From this it follows that the right unit map in $MU_*(MU)$ is as stated in B.4.3.

We can interpret B.4.3 as saying that $MU_*(X)$ is an object in the category $\mathbf{C\Gamma}$ of 3.3, since a coaction of the Hopf algebroid $MU_*(MU)$ is equivalent to an L-module structure with a compatible coaction of the Hopf algebra B, and the latter coaction is equivalent to an action of the group Γ.

B.5 BP-theory

Brown and Peterson [BP66] showed that MU, when localized at a prime p, is homotopy equivalent to an infinite wedge of various suspensions of a 'smaller' spectrum BP (named after them). Its basic properties are

Theorem B.5.1 *Let BP be the Brown-Peterson spectrum described above.*
(a)
$$\pi_*(BP) \cong \mathbf{Z}_{(p)}[v_1, v_2, \ldots]$$
where $\dim v_i = 2p^i - 2$,
(b)
$$H_*(BP) \cong \mathbf{Z}_{(p)}[t_1, t_2, \ldots]$$
where $\dim t_i = 2p^i - 2$, *and*
(c) the generators v_i can be chosen in such a way that the Hurewicz map $h: \pi_*(BP) \to H_*(BP)$ *is given by*

$$h(v_i) = pt_i + decomposables.$$

These v_n's are essentially the same as the ones in 3.3.6 and $\pi_*(BP)$, which we will denote by BP_*, is the same as the ring V of (3.3.8). The role of BP was greatly clarified by the work of Quillen [Qui69], which we will now outline. More details can be found in [Rav86, A2 and 4.1].

Definition B.5.2 *A formal group law F over a torsion free ring R is p-typical if its logarithm has the form*

$$\log_F(x) = \sum_{i \geq 0} \ell_i x^{p^i}.$$

(A more general definition, which does not require R to be torsion free, can be found in [Rav86, A2.1.22].)

There is a universal p-typical formal group law (as in 3.2.3) defined over the ring V of (3.3.8). By Lazard's theorem 3.2.3 it is induced by a homomorphism

$$L \otimes \mathbf{Z}_{(p)} \xrightarrow{\ \tau\ } V \tag{B.5.3}$$

The formal group law over BP^* topologically determined by $BP^*(\mathbf{C}P^\infty)$ is p-typical and the resulting homomorphism $\theta: V \to BP^*$ is an isomorphism.

A theorem of Cartier asserts that every formal group law over a $\mathbf{Z}_{(p)}$-algebra R is a canonically isomorphic to a p-typical one. In the universal example this means there is an idempotent endomorphism

$$L \otimes \mathbf{Z}_{(p)} \xrightarrow{\ \epsilon_p\ } L \otimes \mathbf{Z}_{(p)} \tag{B.5.4}$$

which factors through the τ of (B.5.3). Quillen showed that this homomorphism is induced (in homotopy) by a map of spectra

$$MU_{(p)} \xrightarrow{\epsilon_p} MU_{(p)}$$

called the **Quillen idempotent**. It is characterized by its effect on the coefficients m_i (B.4.2) of the logarithm series, i.e., on the complex projective spaces

$$\epsilon_{p*}(m_i) = \begin{cases} m_i & \text{if } i = p^k - 1 \\ 0 & \text{otherwise.} \end{cases}$$

One then obtains the spectrum BP as the direct limit of the system

$$MU_{(p)} \xrightarrow{\epsilon_p} MU_{(p)} \xrightarrow{\epsilon_p} \cdots.$$

Unfortunately there is no p-typical analog of the group Γ of 3.3.1, i.e., there is no nontrivial collection of power series $\gamma(x)$ such that

$$F^\gamma(x,y) = \gamma(F(\gamma^{-1}(x), \gamma^{-1}(y)))$$

is p-typical whenever F is. For a fixed p-typical F there is a set of power series $\gamma(x)$ with this property, but this set varies with F. This means that the groupoid of p-typical formal group laws over a $\mathbf{Z}_{(p)}$-algebra R and isomorphisms between them is not split in the sense of B.4.4. It follows that the Hopf algebroid $BP_*(BP)$ cannot be constructed from a Hopf algebra over $\mathbf{Z}_{(p)}$ the way $MU_*(MU)$ can be constructed from the Hopf algebra B. We will see an explicit example of this below in (B.5.11).

Thus there is no p-typical analog of the category $\mathbf{C\Gamma}$ of 3.3.2, but there is the category of comodules over $BP_*(BP)$, which we will now describe. For the proof we refer the reader to [Rav86, 4.1 and A1] or [Ada74, II.16]. The result is originally due to Quillen [Qui69].

Theorem B.5.5 (Quillen's theorem) *As a ring,*

$$BP_*(BP) = BP_*[t_1, t_2, \cdots]$$

with $|t_i| = 2p^i - 2$.

Let $\ell_i \in BP_* \otimes \mathbf{Q}$ *denote the image of* m_{p^i-1} *under the Quillen idempotent* ϵ_p *of (B.5.4), with* $\ell_0 = 1$. *The polynomial generators* $v_i \in BP_*$ *are related to the* ℓ_i *recursively by the formula of Araki,*

$$p\ell_n = \sum_{0 \le i \le n} \ell_i v_{n-i}^{p^i} \tag{B.5.6}$$

where it is understood that $v_0 = p$.

The coproduct in $BP_*(BP) \otimes \mathbf{Q}$ is given by

$$\sum_{i,j \geq 0} l_i \Delta(t_j)^{p^i} = \sum_{i,j,k \geq 0} \ell_i t_j^{p^i} \otimes t_k^{p^{i+j}} \tag{B.5.7}$$

where $t_0 = 1$.

The left unit η_L is the standard inclusion

$$BP_* \longrightarrow BP_*(BP) = BP_*[t_1, t_2, \cdots]$$

while the right unit on $BP_*(BP) \otimes \mathbf{Q}$ is given by

$$\sum_{i \geq 0} \eta_R(\ell_i) = \sum_{i,j \geq 0} \ell_i t_j^{p^i}. \tag{B.5.8}$$

We will illustrate the use of these formulae in some simple cases. First observe that in each of (B.5.6), (B.5.7) and (B.5.8), if the sum of the subscripts is n, then the dimension is $2p^n - 2$. For $n = 1$, (B.5.6) gives

$$\begin{aligned} p\ell_1 &= v_1 + p^p \ell_1 & \text{so} \\ \ell_1 &= \frac{v_1}{p - p^p}. \end{aligned}$$

Numbers such as $p - p^p$ occur often in such formulae, so we define

$$\pi_n = p - p^{p^n};$$

each of these is a unit (in $\mathbf{Z}_{(p)}$) multiple of p. Thus we have

$$\ell_1 = \frac{v_1}{\pi_1} \tag{B.5.9}$$

and a similar computation using (B.5.6) for $n = 2$ gives

$$\ell_2 = \frac{v_2}{\pi_2} + \frac{v_1^{p+1}}{\pi_1 \pi_2}.$$

Now we will consider the coproduct formula (B.5.7). For $n = 1$ we get

$$\Delta(t_1) + \ell_1 \Delta(t_0)^p = \ell_1 \otimes 1 + t_1 \otimes 1 + 1 \otimes t_1$$

so

$$\Delta(t_1) = t_1 \otimes 1 + 1 \otimes t_1. \tag{B.5.10}$$

For $n = 2$, (B.5.7) gives

$$\ell_1 \Delta(t_1)^p + \Delta(t_2) = t_2 \otimes 1 + t_1 \otimes t_1^p + 1 \otimes t_2 + \ell_1(t_1^p \otimes 1 + 1 \otimes t_1^p).$$

Using (B.5.9) and (B.5.10) we can rewrite this as

$$
\begin{aligned}
\Delta(t_2) &= t_2 \otimes 1 + t_1 \otimes t_1^p + 1 \otimes t_2 + \ell_1(t_1 \otimes 1 + 1 \otimes t_1^p - \Delta(t_1)^p) \\
&= t_2 \otimes 1 + t_1 \otimes t_1^p + 1 \otimes t_2 - \frac{v_1}{\pi_1} \sum_{0 < i < p} \binom{p}{i} t_1^i \otimes t_1^{p-i}.
\end{aligned}
$$
(B.5.11)

Notice that the coefficient $\binom{p}{i} / \pi_1$ is in $\mathbf{Z}_{(p)}$ because both numerator and denominator are unit multiples of p.

The formulae for Δ and η_R can be restated if we use the following notation. The expression

$$\sum_i {}^F x_i$$

will denote the *formal sum* of the x_i, i.e., it is characterized in a torsion free setting by

$$\log_F \left(\sum_i {}^F x_i \right) = \sum_i \log_F(x_i).$$

Proposition B.5.12 *(a) The coproduct is given by*

$$\sum_{i \geq 0} {}^F \Delta(t_i) = \sum_{i,j \geq 0} {}^F t_i \otimes t_j^{p^i}.$$
(B.5.13)

(b) The right unit is given by

$$\sum_{i,j \geq 0} {}^F v_i t_j^{p^i} = \sum_{i,j \geq 0} {}^F t_i \eta_R(v_j)^{p^i}.$$
(B.5.14)

Proof. (a) (B.5.7) can be rewritten as

$$\sum_{i \geq 0} \log(\Delta(t_i)) = \sum_{i,j \geq 0} \log(t_i \otimes t_j^{p^i})$$

from which the result follows immediately.

(b) (B.5.6) says

$$p \sum_{i \geq 0} \ell_i = \sum_{i,j \geq 0} \ell_i v_j^{p^i}$$

Applying η_R to both sides and using (B.5.8), we get

$$p \sum_{i,j \geq 0} \ell_i t_j^{p^i} = \sum_{i,j,k} \ell_i t_j^{p^i} \eta_R(v_k)^{p^{i+j}}.$$

Using (B.5.6) again to rewrite the left hand side, we get

$$\sum_{i,j,k \geq 0} \ell_i v_j^{p^i} t_k^{p^{i+j}} = \sum_{i,j,k} \ell_i t_j^{p^i} \eta_R(v_k)^{p^{i+j}},$$

which gives the desired formula.   ∎

These formulas demonstrate the integrality of (B.5.7) and (B.5.8) but they are still awkward as stated since they are recursive rather than explicit. In general, explicit formulae in BP-theory are hard to come by in all but the simplest cases, but they are seldom needed for practical purposes. It is usually enough to know the answer modulo some ideal $J \subset BP_*$. *The art of computing with BP-theory is knowing which ideal J to use.* If the chosen J is too large, then one loses the information one wants; if it is too small then the computation is too difficult.

The following approximations are useful.

Proposition B.5.15 *(a) Let*

$$I = (p, v_1, v_2, \cdots) \subset BP_*.$$

Then

$$\Delta(t_n) \equiv \sum_{0 \leq i \leq n} t_i \otimes t_{n-i}^{p^i} \bmod I$$

$$\eta_R(v_n) \equiv \sum_{0 \leq i \leq n} v_i \otimes t_{n-i}^p \bmod I^2.$$

(b) Let

$$I_n = (p, v_1, v_2, \cdots v_{n-1}) \subset BP_*.$$

Then

$$\eta_R(v_n) \equiv v_n \bmod I_n$$

$$\eta_R(v_{n+1}) \equiv v_{n+1} + v_n t_1^{p^n} - v_n^p t_1 \bmod I_n$$

(c) In $BP_*(BP)/(p, v_1, \cdots v_{n-1}, t_1, t_2, \cdots t_{i-1})$ we have

$$\eta_R(v_{n+i}) = v_{n+i} + v_n t_i^{p^n} - v_n^{p^i} t_i.$$

Proof. (a) Reduction mod I converts formal sums to ordinary sums, so the first formula follows immediately from (B.5.13). In (B.5.14) each summand is in I, so we get an ordinary sum by reducing mod I^2. This will kill the terms on the right with $i > 0$, so the equation becomes

$$\sum_{i,j \geq 0} v_i t_j^{p^i} \equiv \sum_{j \geq 0} \eta_R(v_j)$$

as desired.

(b) If we reduce (B.5.14) modulo I_n, all terms in dimension $< |v_n|$ vanish, and there is only one term on each side in that dimension, giving

$$v_n \equiv \eta_R(v_n).$$

This means that there are no cross terms in dimension $|v_{n+1}|$, i.e., we could formally subtract v_n from the left and $\eta_R(v_n)$ from the right, giving

$$v_{n+1} + v_n t_1^{p^n} \equiv \eta_R(v_{n+1}) + t_1 \eta_R(v_n)^p,$$

which gives the desired formula.

(c) If we kill the indicated t_j as well as I_n, then in each dimension below $|v_{n+i}|$ there is at most formal summand on each side of (B.5.14), giving

$$\eta_R(v_{n+j}) = v_{n+j} \qquad \text{for } j < i.$$

These terms can be formally subtracted, giving the desired formula in dimension $|v_{n+i}|$. ∎

The formula for the coproduct $\Delta(t_n)$ above should be compared with the one for $\Delta(\xi_n)$ given in B.3.4. There is a homomorphism

$$BP_*(BP) \longrightarrow A_*$$

induced by the map $BP \to H/(p)$. It sends t_n to $c(\xi_n)$.

Definition B.5.16 *In a Hopf algebroid (S, Σ) (see B.3.7), and ideal $J \subset S$ is **invariant** if $\eta_R(J) \subset J\Sigma$.*

The following result is an easy exercise.

Proposition B.5.17 *If $(S, S \otimes B)$ is a split Hopf algebroid over K (see B.4.5) then an ideal $J \subset S$ is invariant in the sense of B.5.16 if and only if it is invariant under the action of the group $\text{Hom}(B, K)$.*

The ideals I_n are analogous to the ideals $I_{p,n}$ of 3.3.6. One can deduce the following analog of 3.3.6 from B.5.15(c).

Theorem B.5.18 *The only invariant prime ideals in BP_* are the I_n of B.5.15(b), for $0 < n \leq \infty$.*

We will outline the proof. From B.5.15(c) for $n = 0$ we get

$$\eta_R(v_i) \equiv v_i + (p - p^{p^i})t_i \bmod (t_1, t_2, \cdots t_{i-1}).$$

From this we can deduce that the only invariant principal ideals in BP_* are the (p^j), so the only invariant principal prime ideal is (p).

Arguing by induction on n we can show that the only invariant ideals of the form

$$(p, v_1, \cdots v_{n-1}, x)$$

are

$$(p, v_1, \cdots v_{n-1}, v_n^j),$$

and this is prime only when $j = 1$.

There is an analog of the Landweber filtration theorem 3.3.7 which says that every comodule M over $BP_*(BP)$ which is finitely presented as a BP_*-module has a finite filtration

$$0 = F_0 \subset F_1 \subset \cdots F_k = M \tag{B.5.19}$$

such that each subquotient F_i/F_{i-1} is a suspension of BP_*/I_n for some n depending on i.

B.6   The Landweber exact functor theorem

A useful consequence of the Landweber filtration theorem is the Landweber exact functor theorem [Lan76], which we will now describe. It can be stated either in the context of MU-theory or BP-theory; we will do the latter. The question is this: given an BP_*-module M, is the functor

$$X \mapsto BP_*(X) \otimes_{BP_*} M$$

a homology theory (A.3.3)? For the answer to be affirmative, the functor must satisfy the appropriate axioms of A.3.1. The difficult part of this requirement is the exactness axiom. In general, tensoring an exact sequence of BP_*-modules with M does not preserve exactness unless M happens to be flat. Instead, a short exact sequence after tensoring with M leads to a long exact sequence of Tor groups. However, all that we really need is that

$$\mathrm{Tor}_1^{BP_*}(M, BP_*(X)) = 0$$

for any finite complex X. Given the Landweber filtration theorem, this reduces to showing that

$$\text{Tor}_1^{BP_*}(BP_*/I_n, BP_*(X)) = 0 \qquad (\text{B.6.1})$$

for all n.

We can compute this group for $n = 1$ by tensoring M with the short exact sequence

$$0 \longrightarrow BP_* \xrightarrow{\ p\ } BP_* \longrightarrow BP_*/(p) \longrightarrow 0.$$

This gives a 4-term sequence

$$0 \longrightarrow \text{Tor}_1^{BP_*}(M, BP_*/(p)) \longrightarrow M \xrightarrow{\ p\ } M \longrightarrow M/(p) \longrightarrow 0.$$

Thus we see that $\text{Tor}_1^{BP_*}(M, BP_*/(p)) = 0$ provided that M is torsion free, i.e., that multiplication by p in M is monic.

To verify (B.6.1) for general n, we assume inductively that

$$\text{Tor}_1^{BP_*}(M, BP_*/I_{n-1}) = 0.$$

We tensor the short exact sequence

$$0 \longrightarrow \Sigma^{|v_n|} BP_*/I_{n-1} \xrightarrow{\ v_n\ } BP_*/I_{n-1} \longrightarrow BP_*/I_n \longrightarrow 0.$$

with M and get a 4-term sequence

$$0 \longrightarrow \text{Tor}_1^{BP_*}(M, BP_*/I_n) \longrightarrow M/I_{n-1} \xrightarrow{\ p\ } M/I_{n-1} \longrightarrow M/I_n \longrightarrow 0.$$

Thus we see that $\text{Tor}_1^{BP_*}(M, BP_*/I_n) = 0$ provided that M/I_{n-1} is torsion free, i.e., that multiplication by v_n in M/I_{n-1} is monic.

Theorem B.6.2 (Landweber exact functor theorem) *(a) The functor*

$$X \mapsto MU_*(X) \otimes_{MU_*} M$$

for a fixed MU_*-module M is a generalized homology theory if and only for each prime p and positive integer n, multiplication by v_n in

$$M \otimes_{MU_*} MU_*/I_{p,n}$$

is a monomorphism. In particular for such an M there is an MU-module spectrum (see A.2.8) E with $\pi_*(E) = M$.
(b) The functor

$$X \mapsto BP_*(X) \otimes_{BP_*} M$$

*for a fixed BP_*-module M is a generalized homology theory if and only for
each prime p and positive integer n, multiplication by v_n in*

$$M \otimes_{BP_*} BP_*/I_n$$

*is a monomorphism. In particular for such an M there is an BP-module
spectrum (see A.2.8) E with $\pi_*(E) = M$.*

Notice that if M is also a $\mathbf{Z}_{(p)}$-module then the condition is trivially
satisfied for primes q other than p since

$$M \otimes_{MU_*} MU_*/(q) = 0.$$

The condition in the theorem is satisfied by BP_*, regarded as an MU_*-
module via the homomorphism r of (B.5.3), giving another proof of the
existence of the spectrum BP. It is also satisfied by $v_n^{-1}MU_{(p)*}$, by the
BP_*-modules $v_n^{-1}BP_*$, and by

$$E(n)_* = \mathbf{Z}_{(p)}[v_1, v_2, \ldots v_n, v_n^{-1}].$$

In both cases the mod $I_{p,n+1}$ reduction is trivial.

On the other hand the modules

$$
\begin{aligned}
BP\langle n\rangle_* &= \mathbf{Z}_{(p)}[v_1, \ldots v_n], \\
P(n)_* &= BP_*/I_n, \\
B(n)_* &= v_n^{-1}BP_*/I_n, \qquad \text{and} \\
K(n)_* &= \mathbf{Z}/(p)[v_n, v_n^{-1}]
\end{aligned}
$$

do *not* satisfy Landweber's condition. Each of these modules is the coefficient ring of a homology theory, but these functors cannot be described
simply in terms of tensor products as above, e.g. in general

$$K(n)_*(X) \neq MU_*(X) \otimes_{MU_*} K(n)_*.$$

Instead there is a universal coefficient spectral sequence, which generalizes
the universal coefficient theorem expressing $H_*(X; G)$ (for an abelian group
G) in terms of $H_*(X; \mathbf{Z})$. It converges to $K(n)_*(X)$ and its E_2-term is

$$\text{Tor}^{MU_*}(MU_*(X), K(n)_*).$$

There are similar spectral sequences for $BP\langle n\rangle_*(X)$ and $P(n)_*(X)$.

B.7   Morava K-theories

As remarked above, the Morava K-theories *cannot* be constructed by simple use of the Landweber exact functor theorem. However there is another way to construct a BP-module spectrum (see A.2.8) $P(n)$ with $\pi_*(P(n)) = BP_*/I_n$. Würgler ([Wur77] and [Wur86]) showed that it is a ring spectrum with nice multiplicative properties. More precisely,

Theorem B.7.1 *For each prime p and each integer $n > 0$ there is a BP-module spectrum $P(n)$ with*

(i) $\pi_*(P(n)) = BP_*/I_n$;

(ii) *$P(n)$ is a ring spectrum. For $p > 2$ the multiplication is unique and commutative. For $p = 2$ there are two noncommutative multiplications m_1 and m_2 which are opposite to each other, i.e., the diagram*

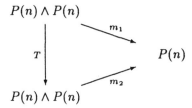

commutes, where T is the map which transposes the two factors;

(iii) *$P(n)$ is flat (A.2.9), i.e. $P(n) \wedge P(n)$ is a wedge of suspensions of $P(n)$; and*

(iv) *for $p > 2$, as a ring*

$$P(n)_*(P(n)) = BP_*(BP)/I_n \otimes E(\tau_0, \tau_1, \cdots \tau_{n-1}) \qquad \text{(B.7.2)}$$

with $|\tau_i| = 2p^i - 1$. For $p = 2$, (B.7.2) is an isomorphism of modules over $BP_*(BP)$ and comodules over $E(\tau_0, \tau_1, \cdots \tau_{n-1})$, but $\tau_i^2 = t_{i+1}$. For all primes the coproduct on the first factor is induced by the one in $BP_*(BP)$ and

$$\Delta(\tau_i) = 1 \otimes \tau_i + \sum_{0 \leq j \leq i} \tau_j \otimes t_{i-j}^{p^j}.$$

The notation τ_i here is slightly misleading since this elements maps to the *conjugate* of τ in the dual Steenrod algebra.

The homology theory $P(n)_*$ enjoys many properties similar to that of BP_*. Let

$$I_m = (v_n, v_{n+1}, \cdots v_{m-1}) \subset P(n)_*$$

for $m \geq n$. These are the only invariant prime ideals in $P(n)_*$. There are analogs of the Landweber filtration theorem and the Landweber exact functor theorem; the latter was proved by Yagita in [Yag76]. It says that if M is a module over $P(n)_* = BP_*/I_n$ then the functor

$$X \mapsto P(n)_*(X) \otimes_{P(n)_*} M$$

is a generalized homology theory if M satisfies certain conditions, namely if for each $m \geq n$, multiplication by v_m induces a monomorphism in

$$M \otimes_{P(n)_*} BP_*/I_m.$$

Now $K(n)_*$ can be regarded as a module over $P(n)_*$ in an obvious way, and it satisfies the conditions. Morava (unpublished) first showed that the functor

$$X \mapsto P(n)_*(X) \otimes_{P(n)_*} K(n)_*$$

is a generalized homology theory, i.e., its satisfies the exactness axiom (A.3.1(b)). This is also proved in [JW75].

$K(n)_*$ is a *graded field* in the sense that every graded module over it is free. One also has

$$K(n) \wedge X \cong \bigvee_\alpha \Sigma^{d(\alpha)} K(n)$$

for any spectrum X. (Recall that a similar statement holds if we replace $K(n)$ by either the rational or the mod p Eilenberg-Mac Lane spectrum.)

This leads to a Künneth isomorphism

$$K(n)_*(X \times Y) = K(n)_*(X) \otimes_{K(n)_*} K(n)_*(Y). \tag{B.7.3}$$

This makes Morava K-theory much easier to compute with than BP-theory. For examples of such computations, see [Rav82], [RW80], [Yam88], [Kuh87], [Hun90], [HKR], [HKR92], and [Ravb]. A corollary of the nilpotence theorem says that the Morava K-theories, along with ordinary homology with field coefficients, are essentially the *only* homology theories with Künneth isomorphisms.

The analog of B.7.1 is

Theorem B.7.4 *For each prime p and each integer $n > 0$ there is a BP-module spectrum $K(n)$ with*

(i) $\pi_*(K(0)) = \mathbf{Q}$ *and* $\pi_*(K(n)) = \mathbf{Z}/(p)[v_n, v_n^{-1}]$;

(ii) $K(n)$ *is a ring spectrum. For $p > 2$ the multiplication is unique and commutative. For $p = 2$ there are two noncommutative multiplications m_1 and m_2 which are opposite to each other, i.e., the diagram*

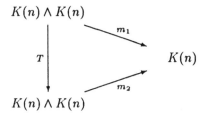

commutes, where T is the map which transposes the two factors;

(iii) $K(n)$ is flat (A.2.9), i.e. $K(n) \wedge K(n)$ is a wedge of suspensions of $K(n)$; and

(iv) for $p > 2$, as a ring

$$K(n)_*(K(n)) = \Sigma(n) \otimes E(\tau_0, \tau_1, \cdots \tau_{n-1}) \qquad \text{(B.7.5)}$$

with $|\tau_i| = 2p^i - 1$ and

$$
\begin{aligned}
\Sigma(n) &= K(n)_* \otimes_{BP_*} BP_*(BP) \otimes_{BP_*} K(n)_* \\
&= K(n)_*[t_1, t_2, \cdots]/(t_i^{p^n} - v_n^{p^i - 1} t_i). \qquad \text{(B.7.6)}
\end{aligned}
$$

For $p = 2$, (B.7.5) is an isomorphism of modules over $BP_*(BP)$ and comodules over $E(\tau_0, \tau_1, \cdots \tau_{n-1})$, but $\tau_i^2 = t_{i+1}$. For all primes the coproduct on the first factor is induced by the one in $BP_*(BP)$, and

$$\Delta(\tau_i) = 1 \otimes \tau_i + \sum_{0 \le j \le i} \tau_j \otimes t_{i-j}^{p^j}.$$

We will explain the multiplicative structure of $\Sigma(n)$. We have

$$K(n)_* \otimes_{BP_*} BP_*(BP) = K(n)_*[t_1, t_2, \cdots],$$

i.e., tensoring on the left kills all of the v_is (including $v_0 = p$) except v_n. It is more difficult to analyze the effect of tensoring on the right with $K(n)_*$. This has the effect of killing $\eta_R(v_i)$ for $i \ne n$. In order to make this more explicit we use the right unit formula (B.5.14). If we set $v_i = 0$ and $\eta_R(v_i) = 0$ for $i \neq n$, then (B.5.14) becomes

$$\sum_{j \geq 0}{}^F v_n t_j^{p^n} = \sum_{i \geq 0}{}^F t_i \eta_R(v_n)^{p^i}$$
$$= \sum_{i \geq 0}{}^F v_n^{p^i} t_i,$$

and this determines the structure of $\Sigma(n)$.

Each side of this equation is a formal sum with at most one summand in each dimension. This means that we can argue by induction on dimension and equate each summand on the left with the corresponding one on the right. This gives

$$v_n t_i^{p^n} = v_n^{p^i} t_i \qquad \text{for } i > 0,$$

as indicated in (B.7.6).

B.8   The change-of-rings isomorphism and the chromatic spectral sequence

In this section we introduce the computational mainspring of this theory, the chromatic spectral sequence. On the algebraic level it is a procedure for getting information about the E_2-term of the Adams-Novikov spectral sequence, i.e., the BP-based Adams spectral sequence (see A.6) converging to $\pi_*(X)$. This E_2 term is

$$\text{Ext}_{BP_*(BP)}(BP_*, BP_*(X)).$$

Here $BP_*(X)$ is a comodule over the Hopf algebroid $BP_*(BP)$, as explained in B.3. In general if M is such a comodule, we will use the notation

$$\text{Ext}(M) = \text{Ext}_{BP_*(BP)}(BP_*, M). \tag{B.8.1}$$

The first thing we need is the *chromatic resolution*. It is a long exact sequence of $BP_*(BP)$-comodules of the form

$$0 \longrightarrow BP_* \longrightarrow M^0 \longrightarrow M^1 \longrightarrow \cdots. \tag{B.8.2}$$

It is a straightforward exercise in homological algebra [Rav86, A1.3.2] to show that there is a spectral sequence converging to $\text{Ext}(BP_*)$ with

$$E_1^{n,s} = \text{Ext}^s(M^n) \quad \text{with} \quad d_r : E_r^{n,s} \to E_r^{n+r,s+1-r}.$$

This is the *chromatic spectral sequence.*

It is also clear that if $BP_*(X)$ is flat over BP_*, then we can tensor it with everything in sight and get a chromatic spectral sequence converging to $\text{Ext}(BP_*(X))$.

(B.8.2) is obtained by splicing together short exact sequences

$$0 \longrightarrow N^n \longrightarrow M^n \longrightarrow N^{n+1} \longrightarrow 0. \qquad (B.8.3)$$

These are defined inductively, starting with

$$
\begin{aligned}
N^0 &= BP_*, \\
M^0 &= BP_* \otimes \mathbf{Q} = p^{-1}N^0 \quad \text{and} \\
N^1 &= BP_* \otimes \mathbf{Q}/\mathbf{Z}_{(p)} = BP_*/(p^\infty).
\end{aligned}
$$

We use the abusive notation $BP_*/(p^\infty)$ because the module can also be described as

$$\varinjlim_i BP_*/(p^i).$$

For $n > 0$ we define

$$M^n = v_n^{-1}N^n,$$

and using the same abusive notation we can write

$$
\begin{aligned}
N^n &= BP_*/(p^\infty, v_1^\infty, \cdots v_{n-1}^\infty) \quad \text{and} &(B.8.4) \\
M^n &= v_n^{-1}BP_*/(p^\infty, v_1^\infty, \cdots v_{n-1}^\infty) &(B.8.5)
\end{aligned}
$$

One has to check that these are actually comodules over $BP_*(BP)$; this done in [Rav86, 5.1.6]. In general if M is a comodule, $v_n^{-1}M$ need not be. The special property of N^n that makes $v_n^{-1}N^n$ a comodule is that each element in N^n is annihilated by some power of the invariant prime ideal I_n (even though N^n as a whole is not).

Each M^n is p-divisible, so for $n > 0$ there is a short exact sequence of comodules

$$0 \longrightarrow M_1^{n-1} \longrightarrow M^n \xrightarrow{\ p\ } M^n \longrightarrow 0$$

where

$$M_1^{n-1} = v_n^{-1}BP_*/(p, v_1^\infty, \cdots v_{n-1}^\infty)$$

with no exponent over the p. M_1^{n-1} is the subgroup of exponent p in M^n.

This short exact sequence leads to a long exact sequence of Ext groups and a Bockstein spectral sequence. This means that in principle we can derive $\text{Ext}(M^n)$ from $\text{Ext}(M_1^{n-1})$.

In M_1^{n-1} multiplication by v_1 is a comodule map, so we have another short exact sequence of comodules

$$0 \longrightarrow M_2^{n-2} \longrightarrow \Sigma^{2p-2} M_1^{n-1} \xrightarrow{v_1} M_1^{n-1} \longrightarrow 0$$

More generally we have

$$0 \longrightarrow M_{i+1}^{n-i-1} \longrightarrow \Sigma^{2p^i-2} M_i^{n-i} \xrightarrow{v_i} M_i^{n-i} \longrightarrow 0 \qquad (\text{B.8.6})$$

for $0 < i < n$, where

$$M_i^{n-i} = v_n^{-1} BP_*/(p, v_1, \cdots v_{i-1}, v_i^\infty, \cdots v_{n-1}^\infty).$$

This means that we can get from

$$\text{Ext}(M_n^0) \qquad \text{to} \qquad \text{Ext}(M^n)$$

by a sequence of n Bockstein spectral sequences. Note that

$$M_n^0 = v_n^{-1} BP_*/I_n.$$

In particular we have

Corollary B.8.7 *If $BP_*(X)$ is a free BP_*-module and there is an integer s_0 such that*

$$\text{Ext}^s(v_n^{-1} BP_*(X)/I_n) = 0 \quad for \quad s > s_0$$

then

$$\text{Ext}^s(BP_*(X) \otimes M^n) = 0 \quad for \quad s > s_0.$$

The group $\text{Ext}(v_n^{-1} BP_*(X)/I_n)$ is especially amenable to computation in view of the following result. *It was this computability that motivated the study leading to the results of this book.* The original insight behind this theorem is due to Jack Morava, and it is related to the constructions described in Chapter 4. The first published proof was given in [MR77]; see also [Rav86, 6.1].

Theorem B.8.8 (Change-of-rings isomorphism) *Let X be a spectrum with $BP_*(X)$ a free BP_*-module. Then there is a natural isomorphism (using the shorthand of (B.8.1))*

$$\text{Ext}(v_n^{-1} BP_*(X)/I_n) = \text{Ext}_{\Sigma(n)}(K(n)_*, K(n)_*(X))$$

where $\Sigma(n)$ is the Hopf algebra of (B.7.5). (The restriction on X implies that

$$K(n)_*(X) = BP_*(X) \otimes_{BP_*} K(n)_*.)$$

Moreover, this Ext group is a module over $K(n)_*$ and

$$\text{Ext}_{\Sigma(n)}(K(n)_*, K(n)_*(X)) \otimes_{K(n)_*} \mathbf{F}_{p^n} = \text{Ext}_{S(n)}(\mathbf{F}_{p^n}, K(n)_*(X) \otimes_{K(n)_*} \mathbf{F}_{p^n})$$

where $S(n)$ is as in (4.2.6), i.e., the ring of continuous $\mathbf{F}_{p^n}$-valued functions on the Morava stabilizer group S_n of 4.2.

This last Ext group can be identified with the cohomology of the group S_n with coefficients in $K(n)_*(X) \otimes_{K(n)_*} \mathbf{F}_{p^n}$. Some cohomological properties of S_n are discussed in 4.3.

Appendix C

Some idempotents associated with the symmetric group

In this appendix we will give the technical details about the representations of the symmetric group needed in 6.4 and 8.3.6. The original source for this material is [Smi]. We are grateful to Jeff Smith for explaining his work to us and to Samuel Gitler for many helpful conversations.

C.1 Constructing the idempotents

In general, the connection between idempotents and representations is as follows. Given a finite group G, a field K and an idempotent $e \in K[G]$, the ideal $(e) \subset K[G]$ is K-vector space with G-action, i.e., a representation of G over K. If K is algebraically closed and its characteristic is prime to the order of G, then it is known that every irreducible representation can be obtained in this way. When the characteristic of K does divide the order of G, the situation is far more complicated even if K is algebraically closed; this is the subject of modular representation theory.

However our goal here is far more modest. We merely need to construct some idempotent elements with certain properties in the p-local group ring $\mathbf{Z}_{(p)}[\Sigma_k]$, where Σ_k is the symmetric group on k letters. We begin by recalling some classical constructions. More details can be found in [Wey39, Chapter IV], [Jam78], [JK81] and many other books on representation theory.

Consider the simplest case first: K is an algebraically closed field and its characteristic does not divide the order of the finite group G. Then the irreducible representations of G over K are in (unnatural) one-to-one correspondence with the conjugacy classes of elements of G. When G is the symmetric group Σ_k, these conjugacy classes are in one-to-one correspondence with the partitions of k. For each partition P of k, we can construct an idempotent in $\mathbf{Q}[\Sigma_k]$ as follows. Suppose we have

$$k = k_1 + k_2 + \cdots k_m \quad \text{with} \quad k_1 \geq k_2 \geq \cdots k_m > 0.$$

We associate to this a *Young diagram*, which is an arrangement of k boxes in m rows, with k_i boxes in the ith row. The boxes may be labelled by the integers from 1 to k, and the left ends of the m rows are vertically aligned.

For example if $k = 12$ and the partition is

$$12 = 5 + 4 + 2 + 1$$

then the Young diagram is

1	2	3	4	5
6	7	8	9	
10	11			
12				

(C.1.1)

The symmetric group Σ_k acts by permuting the boxes (or the labels therein) in the diagram. Let $\Sigma_R \subset \Sigma_k$ denote the subgroup of permutations leaving the rows invariant, so

$$\Sigma_R \cong \Sigma_{k_1} \times \Sigma_{k_2} \times \cdots \Sigma_{k_m}.$$

Let $\Sigma_C \subset \Sigma_k$ denote the subgroup of permutations leaving the columns invariant. For $\sigma \in \Sigma_k$ let $(-1)^\sigma$ denote the sign of σ.

Now define

$$\tilde{e}_P = \sum_{\substack{r \in \Sigma_R \\ c \in \Sigma_C}} (-1)^c rc \quad \in \quad \mathbf{Z}[\Sigma_k].$$

In [Wey39, Lemma 4.3.A] is it shown that $\tilde{e}_P^2 = \mu_P \tilde{e}_P$ for some integer μ_P. If $\mu_P \neq 0$ then

$$e_P = \frac{\tilde{e}_P}{\mu_P} \in \mathbf{Q}[\Sigma_k]$$

is an idempotent. [Wey39, Theorem 4.3.E] asserts that

$$\mu_P g_P = k!$$

where g_P is degree of the representation associated with e_P. A formula for g_P, and hence for μ_P can be found in [Jam78, 20.1] and [JK81, 2.3.21]. Before stating it we need the following definition.

Definition C.1.2 *Given a Young diagram, the* **hook length** $h(i, j)$ *associated with the jth box in the ith row is one more than the sum of the number of boxes to the right of, and the number of boxes below the given box.*

For example, the following table shows the hook lengths of the boxes in the Young diagram of (C.1.1).

8	6	4	3	1
6	4	2	1	
3	1			
1				

Theorem C.1.3 *Let P be the partition of k given by*

$$k = k_1 + k_2 + \cdots k_m \quad \text{with} \quad k_1 \geq k_2 \geq \cdots k_m > 0.$$

Then the integer μ_P defined above is

$$\mu_P = \prod h(i, j) = \frac{\displaystyle\prod_{1 \leq i \leq m} \ell_i!}{\displaystyle\prod_{1 \leq i < j \leq m} (\ell_i - \ell_j)}$$

where $\ell_i = k_i + m - i$.

If we can choose the partition P so that μ_P is not divisible by p, then the resulting idempotent e_P will lie in the p-local group ring $\mathbf{Z}_{(p)}[\Sigma_k]$. For each positive m, consider the partition P given by

$$k_i = (m + 1 - i)(p - 1) \quad \text{and} \quad k = (p - 1)\binom{m+1}{2}.$$

Lemma C.1.4 *For the partition P defined above, the integer μ_P of C.1.3 is not divisible by p.*

Proof. We have

$$
\begin{aligned}
\ell_i &= k_i + m - i \\
&= (p - 1)(m + 1 - i) + m - i \\
&= p(m - i) + p - 1
\end{aligned}
$$

so the second formula of C.1.3 gives

$$
\mu_P = \frac{\displaystyle\prod_{1 \le i \le m} \ell_i!}{\displaystyle\prod_{1 \le i < j \le m} (\ell_i - \ell_j)}
$$

$$
= \frac{\displaystyle\prod_{1 \le i \le m} (p(m-i) + p - 1)!}{\displaystyle\prod_{1 \le i < j \le m} p(j-i)}
$$

If we omit the factors of the numerator not divisible by p, we are left with

$$
\prod_{1 \le i \le m} \left(\prod_{1 \le j \le m-i} pj \right) = \prod_{1 \le i \le m} \left(\prod_{i+1 \le j \le m} p(j-i) \right)
$$

$$
= \prod_{1 \le i < j \le m} p(j-i),
$$

which is precisely the denominator of μ_P. ∎

We will denote the resulting idempotent in $\mathbf{Z}_{(p)}[\Sigma_k]$ by e_m.

Theorem C.1.5 *Let $e_m \in \mathbf{Z}_{(p)}[\Sigma_k]$ as above, let V be a finite dimensional vector space over a field of characteristic p. Let Σ_k act on $W = V^{\otimes k}$ by permuting the factors. Then $e_m W$ is nontrivial if and only if the dimension of V is at least m.*

Proof. We will prove that $e_m W$ is nontrivial for large V by producing a nontrivial vector in it. Let

$$
\{v_1, v_2, v_3, \cdots\}
$$

be a basis of V. Then one has a corresponding basis of W. One of the basis vectors is

$$
w = v_1^{\otimes k_1} \otimes v_2^{\otimes k_2} \otimes \cdots v_m^{\otimes k_m}.
$$

(Recall that $k_i = (p-1)(m+1-i)$.)

To see how e_m acts on a basis vector of W, consider the corresponding Young diagram in which the integers from 1 to k are replaced by basis vectors of V. For w this diagram has v_i in each box in the ith row for each i. For example, when $p = 3$ and $m = 3$, this diagram is

v_1	v_1	v_1	v_1	v_1	v_1
v_2	v_2	v_2	v_2		
v_3	v_3				

Now let w' be the basis vector corresponding to the diagram obtained from that of w by reversing the order of the vectors occurring within each column. In our example, its diagram is

v_3	v_3	v_2	v_2	v_1	v_1
v_2	v_2	v_1	v_1		
v_1	v_1				

Now consider the vector $e_m(w')$. Notice first that no nontrivial element of Σ_C (the group of column-preserving permutations) fixes w'. Moreover, since no basis vector of V appears p times in any given row, the subgroup of Σ_R fixing w' has order prime to p. Hence if we express $e_m(w')$ as a linear combination of basis vectors, the coefficient of w' in this expression will be nontrivial. It follows that $e_m(w')$ is the desired nontrivial vector in $e_m W$.

On the other hand, if the dimension of V is less than m, then each basis vector x of W will have a Young diagram in which some basis vector of V appears at least p times in the top row. It follows that p divides the order of the subgroup of Σ_R fixing x, and therefore e_m annihilates x. ∎

C.2 Idempotents for graded vector spaces

Now we need to discuss the role of signs in the case when V is a graded vector space over a field of odd characteristic. Let

$$V = V^+ \oplus V^-$$

with V^+ concentrated in even dimensions and V^- in odd dimensions. The action of Σ_k on $V^{\otimes k}$ is subject to the usual signs, i.e., a minus sign is introduced each time two vectors in V^- are interchanged.

This means that to get a nontrivial vector in $eV^{\otimes k}$ as in the proof of C.1.5, we need a Young diagram labelled by basis vectors of V in which

- the basis vectors within each column are distinct,

- within each row, no basis vector of V^+ appears p times, and

- no basis vector of V^- appears twice in any row.

With this in mind, let

$$
m_V = \dim V^+ + \left\lceil \frac{\dim V^-}{p-1} \right\rceil
$$

$$
k_V = (p-1) \binom{m_V + 1}{2}
$$

$$
e_V = e_{m_V} \in \mathbf{Z}_{(p)}[\Sigma_{k_V}].
$$

Theorem C.2.1 *With notation as above, let $e_{m_V} \in \mathbf{Z}_{(p)}[\Sigma_{k_V}]$ be as in C.1.5 and $W = V^{\otimes k_V}$. Then $e_V W$ is nontrivial.*

Moreover if $U \subset V$ is a subspace with

$$
\dim U^+ \leq \dim V^+ - 1 \qquad or
$$
$$
\dim U^- \leq \dim V^- - (p-1)
$$

then $e_V U^{\otimes k_V}$ is trivial.

Proof. The argument is similar to that of C.1.5. Let $\{\alpha_i\}$ and $\{\beta_j\}$ be bases of V^+ and V^-. Then let $w \in W$ correspond to the Young diagram labelled as follows. Label each box in the ith row with α_i until the α_i are exhausted. Then use $p-1$ of the β_j in each subsequent row so that there is no repetition within each successive block of $p-1$ columns.

Suppose for example that $p = 3$, $\dim V^+ = 1$ and $\dim V^- = 5$. Then $m_V = 3$ and $k_V = 12$. The Young diagram for w is

α_1	α_1	α_1	α_1	α_1	α_1
β_1	β_2	β_1	β_2		
β_3	β_4				

As before we define a new basis vector w' to be the one associated with the Young diagram obtained from that for w by reversing all of the columns. Thus in our example it is

β_3	β_4	β_1	β_2	α_1	α_1
β_1	β_2	α_1	α_1		
α_1	α_1				

Now in each row, no α_i appears p times and no β_j appears twice. It follows that $e_V(w')$ is nontrivial, so the vector space $e_V W$ is nontrivial.

On the other hand if U is as stated, then

$$\dim U^- + (p-1)\dim U^+ < m_V,$$

so in the diagram for any basis vector u of $U^{\otimes kv}$, either a basis vector of U^- appears more than once or one of U^+ appears at least p times. This means that $e_V(u) = 0$, so $e_V U^{\otimes kv}$ is trivial. ∎

Now suppose that V is a module over either of two Hopf algebras, namely the duals E and T_n of the primitively generated Hopf algebras

$$\begin{aligned} E_* &= E(x) & \text{with } |x| \text{ odd or} \\ T_{n*} &= P(x)/(x^{p^n}) & \text{with } |x| \text{ even and } n > 0. \end{aligned}$$

Then we get a similar module structure on $V^{\otimes m}$ using the Cartan formula.

Theorem C.2.2 *Let V be as in C.1.5 be a module over either E or T_n as defined above, and suppose*

$$V = U \oplus F$$

where F is a nontrivial free module. Then $e_V V^{\otimes kv}$ is a free module over E or T_n.

Proof. First, observe that

$$V^{\otimes kv} = U^{\otimes kv} \oplus F'$$

where F' is free. Now $U^{\otimes kv}$ is invariant under the action of the symmetric group, so we have a short exact sequence

$$0 \longrightarrow e_V U^{\otimes kv} \longrightarrow e_V V^{\otimes kv} \longrightarrow e_V F' \longrightarrow 0.$$

Next observe that the subspace U here satisfies the conditions on U in C.2.1. In the E case, $|x|$ is odd so we have

$$\dim U^+ < \dim V^+ \qquad \text{and} \qquad \dim U^- < \dim V^-.$$

In the T_n case, $|x|$ is even, so either

$$\dim U^+ < \dim V^+ - (p-1) \qquad \text{or} \qquad \dim U^- < \dim V^- - (p-1).$$

It follows that $e_V U^{\otimes kv}$ is trivial and

$$e_V V^{\otimes kv} = e_V F'.$$

This is a summand of the free module F'. E and T_n are both local rings and hence a direct summand of a free module over either of them is free. ∎

C.3   Getting strongly type n spectra from partially type n spectra

In this section we will explain how to use the idempotents discussed above to convert a partially type n spectrum (6.2.5) to a strongly type n spectrum (6.2.3) and thereby (using 6.2.6 and 6.2.4) completing the proof of the periodicity theorem. At the end of the section we will prove the result (C.3.3) for 8.3.6, which is needed in the proof of the smash product theorem (7.5.6).

First we will describe some sub-Hopf algebras of the Steenrod algebra A to which C.2.2 can be applied. V will of course be the mod p cohomology of some spectrum of interest. We want to show how some mild conditions on V can lead to an $e_V V^{\otimes mv}$ on which certain Margolis homology groups vanish.

Recall that Margolis homology groups are defined in terms of the elements P_t^s dual to $\xi_t^{p^s} \in A_*$ for $s < t$ and (for p odd) Q_i dual to τ_i.

Lemma C.3.1 *(i) For each $i \geq 0$, the exterior algebra $E(Q_i)$ is a sub-Hopf algebra of A.*

(ii) For each $n > 0$ the subalgebra generated by P_n^s for $s < n$ is a sub-Hopf algebra isomorphic to the T_n of C.2.2.

Proof. We will give the proof for odd primes only, leaving $p = 2$ (which is easier) as an exercise. Showing that a subalgebra of A is a Hopf algebra is equivalent to showing that its dual is a quotient Hopf algebra of A_*. In the case of $E(Q_i)$, the dual is obtained by setting Q_j for $j \neq i$ and all the ξ_n to zero. Inspection of the coproduct in A_* (B.3.4) shows that the ideal generated by these elements is a Hopf ideal, i.e., it is closed under the coproduct.

More care is required for the subalgebra generated by the P_n^s. It is dual to $P(\xi_n)/(\xi_n^{p^n})$. Hence we need to verify that

$$(\tau_0, \tau_1, \cdots; \xi_1, \cdots, \xi_{n-1}, \xi_n^{p^n}, \xi_{n+1}, \cdots)$$

is a Hopf ideal. Let

$$J_i = (\tau_0, \tau_1, \cdots; \xi_1, \cdots, \xi_{n-1}, \xi_n^{p^n}, \xi_{n+1}, \cdots \xi_{n+i}).$$

It is clear from B.3.4 that J_0 is a Hopf ideal. We will show by induction on i that J_i is as well. Modulo J_{i-1} we have

$$
\begin{aligned}
\Delta(\xi_{n+i}) &= \sum_{0 \leq j \leq n+i} \xi_{n+i-j}^{p^j} \otimes \xi_j \\
&= \xi_{n+i} \otimes 1 + 1 \otimes \xi_{n+i} + \sum_{n \leq j \leq i} \xi_{n+i-j}^{p^j} \otimes \xi_j
\end{aligned}
$$

and the cross terms are each in

$$J_{i-1} \otimes A_* + A_* \otimes J_{i-1}$$

as required. ∎

Theorem C.3.2 *Let X be a partially type n spectrum (6.2.5) and let*

$$V = H^*(X^{(\ell)})$$

for some $\ell > 0$. Then the spectrum

$$Y = e_V(X^{(\ell k v)})$$

is strongly type n (6.2.3) if ℓ is sufficiently large.

Proof. By hypothesis, Q_i for $0 \leq i < n - 1$ and P_t^0 for $1 \leq t \leq n$ act nontrivially on $H^*(X)$. It follows that $H^*(X)^{(\ell)}$ for any $\ell > 0$ contains a summand which is free over $E(Q_i)$.

In the subalgebra T_t of C.3.1(ii), let s_t^i denote the dual of ξ_t^i. Then the Cartan formula gives

$$s_t^i(u \otimes v) = \sum_{0 \leq j \leq i} s_t^j(u) \otimes s_t^{i-j}(v).$$

Now let $x \in H^*(X)$ be a class with $s_t^1(x) \neq 0$. Then for $\ell \geq p^t - 1$, the class

$$s_t^i(x \otimes x \otimes \cdots x) \in H^*(X^{(\ell)})$$

is nontrivial for $0 \leq i \leq p^t - 1$ by the Cartan formula.

It follows that the T_t-submodule of $H^*(X^{(\ell)})$ generated by

$$x \otimes x \otimes \cdots x$$

is free. In order to meet the hypothesis of C.2.2, we need to show that it is a free summand. According to Moore-Peterson [MP73], T_t is self-injective, so a free submodule is always a direct summand.

Hence C.2.2 tells us that $H^*(Y)$ is free over each $E(Q_i)$ and T_t. This means that all of the Margolis homology groups specified in 6.2.3 vanish on Y. ∎

Now we will prove the theorem needed for 8.3.6.

Theorem C.3.3 *Let W be a spectrum where $FK(m)_*(W)$ (with m divisible by $p - 1$; see 4.3) such that the action of every subgroup $H \subset S_m$ of order p is nontrivial, and let*

$$V = FK(m)_*(W^{(p-1)}) = FK(m)_*(W)^{\otimes p-1}.$$

Then the spectrum $e_V W^{(kv(p-1))}$ is such that

$$FK(m)_*(e_V W^{(kv(p-1))}) = e_V FK(m)_*(W^{(kv(p-1))})$$

is a free module over $\mathbf{F}_{p^m}[H]$ for every such H.

Proof. We need to show that for each H, V has a nontrivial free summand over $\mathbf{F}_{p^m}[H]$. Then the result will follow by an argument similar to that of C.2.2.

As a Hopf algebra,

$$\mathbf{F}_{p^m}[H] = \mathbf{F}_{p^m}[x]/(x^p - 1) \quad \text{with} \quad \Delta(x) = x \otimes x.$$

Setting $u = x - 1$, we can write

$$\mathbf{F}_{p^m}[H] = \mathbf{F}_{p^m}[u]/(u^p) \quad \text{with} \quad \Delta(u) = u \otimes 1 + u \otimes u + 1 \otimes u.$$

Since the action of $\mathbf{F}_{p^m}[H]$ on $FK(m)_*(W)$ is nontrivial, we can find nontrivial elements $\alpha, \beta \in FK(m)_*(W)$ with

$$
\begin{aligned}
u(\alpha) &= \beta \\
u(\beta) &= 0,
\end{aligned}
$$

so we have

$$
\begin{aligned}
x(\alpha) &= \alpha + \beta \\
x(\beta) &= \beta \\
x^i(\alpha) &= \alpha + i\beta.
\end{aligned}
\tag{C.3.4}
$$

We will show that

$$v_0 = \alpha^{\otimes(p-1)} \in V$$

generates a free submodule over $\mathbf{F}_{p^m}[H]$ by showing the the vectors $x^i(v_0)$ are linearly independent.

Define vectors $w_j \in V$ for $0 \leq j \leq p - 1$ formally (using t as a dummy variable) by

$$(\alpha + t\beta)^{\otimes(p-1)} = \sum_j t^j w_j,$$

i.e., w_j is the sum of all tensor products with j factors equal to β and $p - 1 - j$ factors equal to α. The w_j are clearly linearly independent.

Then from (C.3.4) we have

$$
\begin{aligned}
x^i(\alpha^{\otimes(p-1)}) &= (\alpha + i\beta)^{\otimes(p-1)} \\
&= \sum_j i^j w_j.
\end{aligned}
$$

These are linearly independent since the Vandermond matrix (i^j) is non-singular, and the result follows. ∎

Bibliography

[AD73]   D. W. Anderson and D. M. Davis. A vanishing theorem in ho-
           mological algebra. *Commentarii Mathematici Helvetici*, 48:318–
           327, 1973.

[Ada66a]   J. F. Adams. On the groups J(X), IV. *Topology*, 5:21–71, 1966.

[Ada66b]   J. F. Adams. A periodicity theorem in homological algebra.
           Proc. Cambridge Phil. Soc., 62:365–377, 1966.

[Ada71]   J. F. Adams. A variant of E. H. Brown's representability theo-
           rem. *Topology*, 10:185–198, 1971.

[Ada74]   J. F. Adams. *Stable Homotopy and Generalised Homology*. Uni-
           versity of Chicago Press, Chicago, 1974.

[Ada75]   J. F. Adams. *Localisation and Completion*. *Lecture Notes in
           Mathematics*, University of Chicago, Department of Mathemat-
           ics, 1975.

[Ada78]   J. F. Adams. *Infinite Loop Spaces. Annals of Mathematics Stud-
           ies*, Princeton University Press, Princeton, 1978.

[AM71]   J. F. Adams and H. R. Margolis. Modules over the Steenrod
           algebra. *Topology*, 10:271–282, 1971.

[BC76]   E. H. Brown and M. Comenetz. Pontrjagin duality for general-
           ized homology and cohomology theories. *American Journal of
           Mathematics*, 98:1–27, 1976.

[BD92]   M. Bendersky and D. M. Davis. 2-primary v_1-periodic homotopy
           groups of $SU(n)$. *American Journal of Mathematics*, 114:465–
           494, 1992.

[BDM]   M. Bendersky, D. M. Davis, and M. Mimura. v_1-periodic ho-
           motopy groups of exceptional Lie groups—torsion-free cases. To
           appear in Transactions of the American Mathematical Society.

[Ben92]   M. Bendersky. The v_1-periodic unstable Novikov spectral se-
           quence. *Topology*, 31:47–64, 1992.

[BF92]   D. J. Benson and M. Feshbach. Stable splittings for the classi-
           fying spaces of finite groups. *Topology*, 31:157–176, 1992.

[BK72]   A. K. Bousfield and D. M. Kan. *Homotopy Limits, Completions
           and Localizations.* Volume 304 of *Lecture Notes in Mathematics*,
           Springer-Verlag, 1972.

[Bou75]   A. K. Bousfield. The localization of spaces with respect to ho-
           mology. *Topology*, 14:133–150, 1975.

[Bou79a]   A. K. Bousfield. The Boolean algebra of spectra. *Commentarii
           Mathematici Helvetici*, 54:368–377, 1979.

[Bou79b]   A. K. Bousfield. The localization of spectra with respect to
           homology. *Topology*, 18:257–281, 1979.

[BP66]   E. H. Brown and F. P. Peterson. A spectrum whose Z_p cohomol-
           ogy is the algebra of reduced p-th powers. *Topology*, 5:149–154,
           1966.

[Bro62]   E. H. Brown. Cohomology theories. *Annals of Mathematics*,
           75:467–484, 1962.

[Car92]   G. Carlsson. A survey of equivariant stable homotpy theory.
           Topology, 31:1–28, 1992.

[CE56]   H. Cartan and S. Eilenberg. *Homological Algebra*. Princeton
           University Press, Princeton, 1956.

[Cec32]   E. Čech. Höherdimensionale homotopiegruppen. In *Verhand-
           lungen des Internationalen Mathematikerkongress, Zürich, 1932*,
           page 203, Orel Füssli, Zürich and Leipzig, 1932.

[CMN79]   F. R. Cohen, J. C. Moore, and J. A. Neisendorfer. Torsion in
           homotopy groups. *Annals of Mathematics*, 109:121–168, 1979.

[Dav91]   D. M. Davis. The v_1-periodic homotopy groups of $SU(n)$ at
           odd primes. *Proceedings of the London Mathematical Society
           (3)*, 43:529–544, 1991.

[Dev]   E. Devinatz. Small ring spectra. To appear.

[Dev90]   E. Devinatz. K-theory and the generating hypothesis. *American Journal of Mathematics*, 112:787–804, 1990.

[DHS88]   E. Devinatz, M. J. Hopkins, and J. H. Smith. Nilpotence and stable homotopy theory. *Annals of Mathematics*, 128:207–242, 1988.

[DM]   D. M. Davis and M. E. Mahowald. Some remarks on v_1-periodic homotopy groups. In N. Ray and G. Walker, editors, *Adams Memorial Symposium on Algebraic Topology Volume 2*, pages 55–272, Cambridge University Press, Cambridge, 1992.

[DMW92] W. G. Dwyer, H. R. Miller, and C. W. Wilkerson. Homotopical uniqueness of classifying spaces. *Topology*, 31:29–46, 1992.

[Dol80]   A. Dold. *Lectures on Algebraic Topology*. Springer-Verlag, New York, 1980.

[ES52]   S. Eilenberg and N. E. Steenrod. *Foundations of Algebraic Topology*. Princeton University Press, Princeton, 1952.

[Fre37]   H. Freudenthal. Über die Klassen der Sphärenabbildungen. *Compositio Math.*, 5:299–314, 1937.

[Fre66]   P. Freyd. Stable homotopy. In S. Eilenberg, D. K. Harrison, S. Mac Lane, and H. Röhrl, editors, *Proceedings of the Conference on Categorical Algebra*, pages 121–172, Springer-Verlag, 1966.

[GH81]   M. J. Greenberg and J. R. Harper. *Algebraic Topology: A First Course*. Benjamin/Cummins, Reading, Massachusetts, 1981.

[Gra75]   B. Gray. *Homotopy Theory: An Introduction to Algebraic Topology*. Academic Press, New York, 1975.

[Haz78]   M. Hazewinkel. *Formal Groups and Applications*. Academic Press, New York, 1978.

[Hig71]   P. J. Higgins. *Categories and Groupoids*. Van Nostrand Reinhold, London, 1971.

[HKR]   M. J. Hopkins, N. J. Kuhn, and D. C. Ravenel. Generalized group characters and complex oriented cohomology theories. Submitted to *Journal of the AMS*.

[HKR92]   M. J. Hopkins, N. J. Kuhn, and D. C. Ravenel. Morava K-theories of classifying spaces and generalized characters for finite groups. In J. Aguadé, M. Castellet, and F. R. Cohen, editors, *Algebraic Topology: Homotopy and Group Cohomology*, pages 186–209, Springer-Verlag, New York, 1992.

[Hop87]   M. J. Hopkins. Global methods in homotopy theory. In J. D. S. Jones and E. Rees, editors, *Proceedings of the 1985 LMS Symposium on Homotopy Theory*, pages 73–96, 1987.

[HR]   M. J. Hopkins and D. C. Ravenel. A proof of the smash product conjecture. To appear.

[HS]   M. J. Hopkins and J. H. Smith. Nilpotence and stable homotopy theory II. To appear.

[Hun90]   J. R. Hunton. The Morava K-theories of wreath products. *Math. Proc. Cambridge Phil. Soc.*, 107:309–318, 1990.

[Hur35]   W. Hurewicz. Beiträge zur Topologie Deformationen I–II. *Nederl. Akad. Wetensch. Proc.*, Series A, 38:112–119, 521–528, 1935.

[Hur36]   W. Hurewicz. Beiträge zur Topologie Deformationen III–IV. *Nederl. Akad. Wetensch. Proc.*, Series A, 39:117–126, 215–224, 1936.

[Jam55]   I. M. James. Reduced product spaces. *Annals of Mathematics*, 62:170–197, 1955.

[Jam78]   G. D. James. *The Representation Theory of the Symmetric Groups*. Volume 682 of *Lecture Notes in Mathematics*, Springer-Verlag, New York, 1978.

[JK81]   G. D. James and A. Kerber. *The Representation Theory of the Symmetric Group*. Volume 16 of *Encyclopedia of Mathematics and its Applicatind its Applications*, Cambridge University Press, Cambridge, 1981.

[JM92]   S. Jackowski and J. McClure. Homotopy decomposition of classifying spaces via elementary abelian groups. *Topology*, 31:113–132, 1992.

[JW75]   D. C. Johnson and W. S. Wilson. BP-operations and Morava's extraordinary K-theories. *Mathematische Zeitschrift*, 144:55–75, 1975.

[JY80]   D. C. Johnson and Z. Yosimura. Torsion in Brown-Peterson homology and Hurewicz homomorphisms. *Osaka Journal of Mathematics*, 17:117–136, 1980.

[Kuh87]   N. J. Kuhn. Morava K-theories of some classifying spaces. *Transactions of the American Mathematical Society*, 304:193–205, 1987.

[Lan67]   P. S. Landweber. Cobordism operations and Hopf algebras.
           Transactions of the American Mathematical Society, 129:94–110,
           1967.

[Lan73a]   P. S. Landweber. Annihilator ideals and primitive elements in
           complex cobordism. *Illinois Journal of Mathematics*, 17:273–
           284, 1973.

[Lan73b]   P. S. Landweber. Associated prime ideals and Hopf algebras.
           Journal of Pure and Applied Algebra, 3:175–179, 1973.

[Lan76]   P. S. Landweber. Homological properties of comodules over
           $MU_*(MU)$ and $BP_*(BP)$. *American Journal of Mathematics*,
           98:591–610, 1976.

[Lan79]   P. S. Landweber. New applications of commutative algebra
           to Brown-Peterson homology. In P. Hoffman and V. Snaith,
           editors, *Algebraic Topology, Waterloo 1978*, pages 449–460,
           Springer-Verlag, New York, 1979.

[Laz55a]   M. Lazard. Lois de groupes et analyseurs. *Ann. Écoles Norm.
           Sup.*, 72:299–400, 1955.

[Laz55b]   M. Lazard. Sur les groupes de Lie formels á une paramètre.
           Bull. Soc. Math. France, 83:251–274, 1955.

[Laz65]   M. Lazard. Groupes analytiques p-adic. *Inst. Hautes Études
           Sci. Publ. Math.*, 26, 1965.

[Lim60]   E. L. Lima. Stable Postnikov invariants and their duals. *Summa
           Brasil. Math.*, 4:193–251, 1960.

[LRS]   P. S. Landweber, D. C. Ravenel, and R. E. Stong. Periodic
           cohomology theories defined by elliptic curves. Submitted to
           Topology.

[Mah77]   M. E. Mahowald. A new infinite family in $_2\pi_*^S$. *Topology*,
           16:249–256, 1977.

[Mah81]   M. E. Mahowald. *bo*-resolutions. *Pacific Journal of Mathematics*, 192:365–383, 1981.

[Mah82]   M. E. Mahowald. The image of J in the EHP sequence. *Annals
           of Mathematics*, 116:65–112, 1982.

[Mar83]   H. R. Margolis. *Spectra and the Steenrod Algebra: Modules over
           the Steenrod Algebra and the Stable Homotopy Category*. North-
           Holland, New York, 1983.

[Mil58] J. W. Milnor. The Steenrod algebra and its dual. *Annals of Mathematics*, 67:150–171, 1958.

[Mil59] J. W. Milnor. On spaces having the homotopy type of CW-complex. *Transactions of the American Mathematical Society*, 90:272–280, 1959.

[Mil60] J. W. Milnor. On the cobordism ring Ω^* and a complex analogue, Part I. *American Journal of Mathematics*, 82:505–521, 1960.

[Mil62] J. W. Milnor. On axiomatic homology theory. *Pacific J. Math.*, 12:337–341, 1962.

[Mil81] H. R. Miller. On relations between Adams spectral sequences, with an application to the stable homotopy of a Moore space. *Journal of Pure and Applied Algebra*, 20:287–312, 1981.

[Mit85] S. A. Mitchell. Finite complexes with $A(n)$-free cohomology. *Topology*, 24:227–248, 1985.

[Mor85] J. Morava. Noetherian localizations of categories of cobordism comodules. *Annals of Mathematics*, 121:1–39, 1985.

[MP73] J. C. Moore and F. P. Peterson. Nearly Frobenius algebras, Poincaré algebras and theori modules. *Journal of Pure and Applied Algebra*, 3:83–93, 1973.

[MP92] J. Martino and S. B. Priddy. The complete stable splitting for the classifying space of a finite group. *Topology*, 31:143–156, 1992.

[MR77] H. R. Miller and D. C. Ravenel. Morava stabilizer algebras and the localization of Novikov's E_2-term. *Duke Mathematical Journal*, 44:433–447, 1977.

[MRW77] H. R. Miller, D. C. Ravenel, and W. S. Wilson. Periodic phenomena in the Adams-Novikov spectral sequence. *Annals of Mathematics*, 106:469–516, 1977.

[MS74] J. W. Milnor and J. D. Stasheff. *Characteristic Classes*. Volume 76 of *Annals of Mathematics Studies*, Princeton University Press, Princeton, 1974.

[MW81] H. R. Miller and C. Wilkerson. Vanishing lines for modules over the Steenrod algebra. *Journal of Pure and Applied Algebra*, 22:293–307, 1981.

[Nis73] G. Nishida. The nilpotence of elements of the stable homotopy groups of spheres. *Journal of the Mathematical Society of Japan*, 25:707–732, 1973.

[Nov60] S. P. Novikov. Some problems in the topology of manifolds connected with the theory of Thom spaces. *Soviet Mathematics Doklady*, 1:717–720, 1960.

[Nov62] S. P. Novikov. Homotopy properties of Thom complexes (Russian). *Mat. Sb. (N. S.)*, 57 (99):407–442, 1962.

[Nov67] S. P. Novikov. The methods of algebraic topology from the viewpoint of cobordism theories. *Math. USSR—Izv.*, 827–913, 1967.

[Qui69] D. G. Quillen. On the formal group laws of oriented and unoriented cobordism theory. *Bulletin of the American Mathematical Society*, 75:1293–1298, 1969.

[Rava] D. C. Ravenel. A counterexample to the telescope conjecture. To appear.

[Ravb] D. C. Ravenel. The homology and Morava K-theory of $\Omega^2 SU(n)$. To appear in *Forum Mathematicum*.

[Rav82] D. C. Ravenel. Morava K-theories and finite groups. In S. Gitler, editor, *Symposium on Algebraic Topology in Honor of José Adem*, pages 289–292, American Mathematical Society, Providence, Rhode Island, 1982.

[Rav84] D. C. Ravenel. Localization with respect to certain periodic homology theories. *American Journal of Mathematics*, 106:351–414, 1984.

[Rav86] D. C. Ravenel. *Complex Cobordism and Stable Homotopy Groups of Spheres*. Academic Press, New York, 1986.

[Rav87] D. C. Ravenel. The geometric realization of the chromatic resolution. In W. Browder, editor, *Algebraic topology and algebraic K-theory*, pages 168–179, 1987.

[Rav92] D. C. Ravenel. Progress report on the telescope conjecture. In N. Ray and G. Walker, editors, *Adams Memorial Symposium on Algebraic Topology Volume 2*, pages 1–21, Cambridge University Press, Cambridge, 1992.

[Rec70]   D. L. Rector. Steenrod operations in the Eilenberg-Moore spec-
          tral sequence. *Commentarii Mathematici Helvetici*, 45:540–552,
          1970.

[RW77]   D. C. Ravenel and W. S. Wilson. The Hopf ring for complex
          cobordism. *Journal of Pure and Applied Algebra*, 9:241–280,
          1977.

[RW80]   D. C. Ravenel and W. S. Wilson. The Morava K-theories of
          Eilenberg-MacLane spaces and the Conner-Floyd conjecture.
          American Journal of Mathematics, 102:691–748, 1980.

[SE62]   N. E. Steenrod and D. B. A. Epstein. *Cohomology Operations.*
          Volume 50 of *Annals of Mathematics Studies*, Princeton Univer-
          sity Press, Princeton, 1962.

[Ser53]   J.-P. Serre. Groupes d'homotopie et classes de groupes abelien.
          Annals of Mathematics, 58:258–294, 1953.

[Ser65]   J.-P. Serre. Sur la dimension cohomologique des groupes profi-
          nis. *Topology*, 3:413–420, 1965.

[Shi86]   K. Shimomura. On the Adams-Novikov spectral sequence
          and products of β-elements. *Hiroshima Mathematical Journal*,
          16:209–224, 1986.

[Smi]   J. Smith. Finite complexes with vanishing lines of small slope.
          To appear.

[Smi69]   L. Smith. On the construction of the Eilenberg-Moore spec-
          tral sequence. *Bulletin of the American Mathematical Society*,
          75:873–878, 1969.

[Smi71]   L. Smith. On realizing complex cobordism modules, IV, Ap-
          plications to the stable homotopy groups of spheres. *American
          Journal of Mathematics*, 99:418–436, 1971.

[Sna74]   V. Snaith. Stable decomposition of $\Omega^n \Sigma^n X$. *Journal of the
          London Mathematical Society*, 7:577–583, 1974.

[Spa66]   E. H. Spanier. *Algebraic Topology.* McGraw-Hill, New York,
          1966.

[ST86]   K. Shimomura and H. Tamura. Non-triviality of some composi-
          tions of β-elements in the stable homotopy of the Moore spaces.
          Hiroshima Mathematical Journal, 16:121–133, 1986.

[Sto68]   R. E. Stong. *Notes on Cobordism Theory.* Princeton University Press, Princeton, 1968.

[Swi75]   R. Switzer. *Algebraic Topology — Homotopy and Homology.* Springer-Verlag, New York, 1975.

[Tho54]   R. Thom. Quelques propriétés globales des variétés differentiables. *Commentarii Mathematici Helvetici*, 28:17–86, 1954.

[Tho90]   R. J. Thompson. Unstable v_1-periodic homotopy at odd primes. *Transactions of the American Mathematical Society*, 319:535–559, 1990.

[Tod60]   H. Toda. On unstable homotopy of spheres and classical groups. *Proceedings of the National Academy of Sciences U.S.A.*, 46:1102–1105, 1960.

[Tod67]   H. Toda. An important relation in the homotopy groups of spheres. *Proceedings of the Japan Academy*, 43:893–942, 1967.

[Tod68]   H. Toda. Extended p-th powers of complexes and applications to homotopy theory. *Proceedings of the Japan Academy*, 44:198–203, 1968.

[Tod71]   H. Toda. On spectra realizing exterior parts of the Steenrod algebra. *Topology*, 10:53–65, 1971.

[Vic73]   J. W. Vick. *Homology Theory: An Introduction to Algebraic Topology.* Academic Press, New York, 1973.

[Wey39]   H. Weyl. *The Classical Groups: Their Invariants and Representations.* Princeton University Press, Princeton, 1939.

[Whi62]   G. W. Whitehead. Generalized homology theories. *Transactions of the American Mathematical Society*, 102:227–283, 1962.

[Whi78]   G. W. Whitehead. *Elements of Homotopy Theory.* Springer-Verlag, New York, 1978.

[Wur77]   U. Würgler. On products in a family of cohomology theories associated to the invariant prime ideals of $\pi_*(BP)$. *Commentarii Mathematici Helvetici*, 52:457–481, 1977.

[Wur86]   U. Würgler. Commutative ring-spectra in characteristic 2. *Commentarii Mathematici Helvetici*, 61:33–45, 1986.

[Yag76]   N. Yagita. The exact functor theorem for BP_*/I_n-theory. *Proceedings of the Japan Academy*, 52:1–3, 1976.

[Yam88]   A. Yamaguchi.   Morava K-theory of double loop spaces of
           spheres. *Mathematische Zeitschrift*, 199:511–523, 1988.

Index